I0006094

Sources of Information Value

Also by the authors

INFORMATION SYSTEMS STRATEGIC PLANNING: A Source of Competitive Advantage *(R. Andreu, Josep Valor-Sabatier and Joan E. Ricart-Costa*

MANAGING STRATEGICALLY IN AN INTERCONNECTED WORLD *(edited by Joan E. Ricart-Costa, Michael A. Hitt and Robert D. Nixon)*

THE VIRTUAL SOCIETY *(Josep Valor-Sabatier, Brian Subirana and Joan E. Ricart-Costa)*

Sources of Information Value

Strategic Framing and the Transformation of the Information Industries

Joan E. Ricart-Costa
Brian Subirana
and
Josep Valor-Sabatier

© Joan E. Ricart-Costa, Brian Subirana and Josep Valor-Sabatier 2004

Foreword © Pankaj Ghemawat 2004

All rights reserved. No reproduction, copy or transmission of this publication may be made without written permission.

No paragraph of this publication may be reproduced, copied or transmitted save with written permission or in accordance with the provisions of the Copyright, Designs and Patents Act 1988, or under the terms of any licence permitting limited copying issued by the Copyright Licensing Agency, 90 Tottenham Court Road, London W1T 4LP.

Any person who does any unauthorized act in relation to this publication may be liable to criminal prosecution and civil claims for damages.

The authors have asserted their rights to be identified as the authors of this work in accordance with the Copyright, Designs and Patents Act 1988.

First published 2004 by
PALGRAVE MACMILLAN
Houndmills, Basingstoke, Hampshire RG21 6XS and
175 Fifth Avenue, New York, N.Y. 10010
Companies and representatives throughout the world

PALGRAVE MACMILLAN is the global academic imprint of the Palgrave Macmillan division of St. Martin's Press, LLC and of Palgrave Macmillan Ltd. Macmillan® is a registered trademark in the United States, United Kingdom and other countries. Palgrave is a registered trademark in the European Union and other countries.

ISBN 978-1-4039-1233-6

This book is printed on paper suitable for recycling and made from fully managed and sustained forest sources.

A catalogue record for this book is available from the Library of Congress

10   9   8   7   6   5   4   3   2   1
13 12 11 10 09 08 07 06 05 04

Typeset by Cambrian Typesetters, Frimley, Camberley, Surrey, England

To our wives: Maria Carme, Mariona and Montse

Contents

Contents

LIST OF FIGURES

LIST OF TABLES

LIST OF EXHIBITS

LIST OF ABBREVIATIONS

ADSL   Assymmetric Digital Subscriber Line
AOL   America Online
ARPU   Average Revenue Per User
ASP   Application Service Provider
BCG   Boston Consulting Group
BTB   Business To Business
BTC   Business To Consumer
CPM   Cost per thousand
CRS   Computer Reservation Service
CTB   Consumer to Business
CTC   Consumer to Consumer
DSL   Ditigal Subscriber Line
DVD   Digital Video Disc
ENIAC   Electronic Numerical integrator and Calculator
ERP   Enterprise Resource Planning
GPRS   General packet Radio Service
GPS   Geographic Position by Satellite
GSM   Global System for Mobile
HTML   Hypertext MarkUp Language
IAP   Internet Access Provider
ICTs   Information and Communication Technologies
IP   Internet Protocol
ISDN   Integrated Service Digital Network
ISP   Internet Service Provider
KDD   Knowledge Discovery in Databases
MPEG   Moving Picture Experts Group
MSN   Microsoft Network
NYT   New York Time
OSP   Online Service Providers
PDA   Personal Digital Assistant
POP   Point of Presence
POTS   Plain Old Telephone Service
P2P   Peer to Peer
RFID   Radio Frequency Identification
SMS   Short Message Servicing
TCP   Transfer Control Protocol
3G   Third Generation of mobile
UMTS   Universal Mobile Telecommunications System
WiFi   Wireless Fidelity
WWW   World Wide Web
XML   Extensible Markup Language

Sources of Information Value is a book dealing with the information industries in a broad sense. It provides a framework of analysis to help understand the different competitive strategies of companies that happen to compete fiercely from within seemingly unrelated industries. These include, among others, telecommunication carriers, software developers and content providers. Our analysis encompasses all business opportunities involved in the process of bringing information to a customer, from the producer of the information item to the company that has written the software to make it available in a suitable form.

To foresee the future of an industry, as well as to forecast whether the most prominent business models are viable, it is essential to differentiate between *value creation* and *value appropriation*. In the high points of the Internet expansion, it was believed that the disruptive nature of this technology had changed the fundamentals of business. A new era of competition in which none of the old paradigms was valid was heralded. The collapse of the technology market and the high-profile failures of many of the upstart dot.coms have shown us that the old business rules still apply. Ebusinesses had shown great value propositions on the value creation side through the reduction of transaction costs, search costs or enhanced customization opportunities. Still, value appropriation appeared to be very problematic. Although new products and pricing mechanisms might help companies in appropriating the created value, recent reality has shown that increased rivalry and constant entrance of new competitors, as well as increased market transparency, pose significant challenges to value appropriation by the firms. Even companies that own significant assets, such as telecommunication carriers, are suffering a seemingly endless reduction in their traditional sources of revenue, forcing them to explore new business avenues that are foreign to their managers and investors alike.

This book systematically analyzes most of the business activities that the Internet has allowed to emerge, both as startups as well as traditional companies who have embraced the new technology, and tries to determine whether there are some business models that might produce superior returns and sustainability. An initial observation that prompted us to study these industries was that many firms ventured themselves into activities not necessarily core to their main activity. Although competing in multiple industries has been a practice that large companies have practiced for many years, including well-known diversified conglomerates like General Electric and more focused firms like Procter and Gamble, we found that traditional models did not explain success or failure in the new information economy.

This book answers these questions by providing a framework inspired by the traditional value chain that we have termed the '*information value chain*'. We will analyze each sector, from content providers, to software manufacturers, to telecommunications operators, and show that each one of these sectors has structural characteristics that makes them very unattractive and rarely profitable in the long run, except when a company dominates a standard. Unfortunately, open standards push entire sectors to commoditization and we have observed extraordinary drops of profitability. Additionally, technology disruptions change the competitive landscape and what seemed sure bets a few years ago, like fiber optics to the home, now are questioned in the face of technologies based on existing copper wires (DSL), or wireless (LDMS and Wi-Fi.)

A whole chapter is devoted to extensions of the value chain, such as transaction streams, that are not easily modeled with a linear step-wise model. These generalize the information value chain framework to more complex environments. Transaction streams are electronic markets where more than one player controls each component of a transaction. Throughout the book, an architecture of transaction streams in web-like networks is constructed, leading to a full-fledge market example based on doubleclick.com, with more than 20 players involved in an apparently single transaction, such as accessing a web page or verifying a purchased good using RFIDs.

We build up our information value chain model by reviewing, in the light of the traditional strategic management models, some of the most successful strategies in information-based industries. Different chapters will examine (1) product bundling, which has been argued to be a very effective strategy for value appropriation of information goods; (2) vertical integration, which has often been claimed to be adequate to boost efficiencies along the overall supply chain of an industry; and (3) system lock-in, which allows the companies that master it to obtain superior market dominance through the use of network externalities. We claim that these models are insufficient to carry out a comprehensive strategic analysis of this broad 'industry' and that to understand these new competitive environments of information goods it is necessary to adopt a wider point of view that takes into account the complementary strategies carried out in different competitive markets.

We call the proposed framework of analysis '*strategic framing*' and the broader bundling strategy '*market bundling*', and show evidence that a bundled market proposition throughout the value system seems to be a superior strategy providing a road to overall success. Additionally, an entire section is devoted to the issue of standards, where we will discuss the struggle to dominate different steps of the value chain by providing interconnectivity, at both the terminal and server levels. We will argue from both sides, the pros and cons of the controversy with open source software.

In the process of developing the framework we will describe and analyze the strategies of the companies we consider paradigms in the information industries and show how market bundling opportunities have affected in different ways the value appropriation possibilities of these key players. Many of these mini-cases are described in boxed inserts alongside the text for easier reviewing.

Our model is just a tool. A tool to re-invent your business. A tool to create a better future for information-based companies. A tool to frame your strategy so that you ask yourself the right questions. The first step to reach the right answers.

Acknowledgments

We owe recognition to a number of people and institutions. IESE Business School in Barcelona has provided us with time to develop our ideas and put them into writing. The PwC-IESE eBusiness Center has been instrumental in financing parts of the research, case studies, and assistant support without which this book would had not seen the light. Extremely interesting conversations with our colleagues Rafael Andreu and Sandra Sieber at IESE have helped us refine our ideas on strategic framing and market bundling. The seeds of this book come from having had the pleasure to work on several occasions with Professors Thomas Malone and Arnoldo Hax from MIT, and Professor Pankaj Gemawhat from Harvard Business School, at their institutions as well as while on sabbatical at IESE. Special thanks go to Robert A. Mamis who helped us immensely with parts of the writing and to our research assistants Carles Cabré, Eduard Guiu, and in particular, Guillermo Armelini who has spent endless hours fetching the last reference, checking details and figures, and coordinating with the editor.

This book, like many other endeavors in our lives, would have not been possible without the continuous encouragement of our wives Maria Carme, Mariona, and Montse.

The authors and publishers are grateful to Nokia, NTT, DoCoMo, Sony-Ericsson, Hoover, Google, and Forrester Research for permission to reprint illustrative material. Every effort has been made to contact all copyright-holders, but if any have been inadvertently omitted the publisher will be pleased to make the necessary arrangement at the earliest opportunity.

The old millennium yielded to the new one not with a bang or a whimper, but with a bubble bursting. With hindsight, hot air as well as the fundamentals helped float an information technology (IT) investment bubble that, when it burst, brought about at least a few years of greater general sobriety about the uses and limits of IT.

What remains controversial is whether the current cooling of interest in IT is simply a cyclical (and natural) correction to the excesses of the late 1990s or whether it also reflects – or should reflect – what is claimed to be the diminishing strategic importance of IT over time. In the best-known version of this controversial claim, IT's increasing potency and ubiquity are supposed to have shrunk its strategic importance by leading to its commoditization and loss of scarcity value (Nicholas G. Carr, 'IT Doesn't Matter', *Harvard Business Review*, May 2003). If true, this would imply a downgrade of IT as a topic for top management's strategic attentions and multiple ripple effects – not least for the authors or readers of books such as this one. So, to set the stage, it may be useful for me to begin by explaining my own sense of the 'IT does/doesn't matter' argument, and then relating this book to it.

My sense is that the recent argument about IT not mattering much strategically any more is about as far off base as some of the more breathless pronouncements from the late 1990s (in some of the same publication outlets) about the brave new world being unlocked by IT. (Why ideas about strategy seem prone to overshoots of this sort is an interesting topic, but one that would take us too far off track.) One counter-argument is suggested by simple arithmetic: it just doesn't seem sensible for a company to allocate 50–60 per cent of its investment capital without a clear sense of how that investment would help create or sustain a competitive advantage. Yet that is what some of the self-described takeaways from the 'IT doesn't matter' argument – spend less; follow, don't lead; focus on vulnerabilities, not opportunities – would entail.

A second counter to the argument about IT not mattering much strategically any more is by way of analogy with other technology revolutions in modern times. Take the case of electricity, which had penetrated less than 10 per cent of US households and factories by 1899. It took the better part of another two decades or longer, depending on the segment, for penetration to cross the 50 per cent mark. For this reason and others, there was a substantial time lag between the basic technical breakthroughs and measurable productivity impacts. When they did come, however, the productivity impacts proved to be large: in the 1920s, the US total factor productivity (TFP) growth rate reached unprecedented levels, with electrification indubitably playing an important role. Even more strikingly, the stockmarket effectively discounted these productivity effects well before they had mostly manifested themselves. Shares in electrical power and equipment companies rose more rapidly than general industrial indexes through much of the 1890s and the first few years of the 1900s, but in 1905 crashed back to early 1890 levels, and failed to recover all the ground they had lost by 1925 – although equipment companies did fare significantly better, on average, than power companies. Of course, there was also important variation around performance averages: General Electric, for instance, built up a global position in electrical lighting more than hundred years ago, that has proved highly profitable and sustainable to the present day.

If the analogy between IT and electricity holds, the assertion early in the twenty-first century that 'IT doesn't matter' would seem about as accurate as an announcement early in the twentieth that the game in electricity was over. Other analogies support similar conclusions. Thus, based on a broad review of episodes of technological and entrepreneurial breakthroughs, economist Bradford de Long ('Old Rules for the New Economy', *Rewired*, September 12, 1997), has proposed that 'In every sector, the heroic period of rapid technological progress comes to an end . . . [but] just because price asymptotes at zero doesn't mean revenue does . . . and some firms wind up larger than anyone had every imagined.' In other words, the vision of new technologies quickly congealing into boring old infrastructure that offers ever less room for building sustainable competitive advantages simply doesn't mesh with past experience. Instead, while prices and costs shrivel, volume expands greatly, and a few firms can end up accounting for a large share of the total value captured by an industry.

A third counter to the argument that 'IT doesn't matter' argument is based on the idea that IT is harder than, say, the railroad network or the power grid, to cast as a generic infrastructure-like resource given how deeply specialized appropriate IT is to firms' histories, organizational arrangements and intended strategies (an early but specific and subtle discussion of historical and organization influences on internal information flows was provided by Kenneth Arrow in his *The Limits of Organization*, 1974). Take the case of Wal-Mart, which in addition to having grown to be the largest firm in the world, has also been characterized (by Robert Solow) as the single biggest factor behind the acceleration of US productivity growth in the second half of the 1990s. A significant fraction of the improvement in labor productivity at Wal-Mart as generated by IT, in which Wal-Mart has invested relatively heavily for decades, writing many applications in-house to ensure interoperability and alignment with business needs. One key result was the company's Retail Link private exchange, one of the first business-to-business e-commerce networks, which provides all of Wal-Mart's suppliers with computer access to point-of-sale (POS) data to analyze sales trends and the inventories of their products on a store-by-store basis. Retail Link reportedly cost Wal-Mart $4 billion to develop and perfect and also required substantial investment by suppliers (by one estimate, an order of magnitude more) to be implemented. But as a result, by 2002, it took less than 10 minutes for information captured by POS scanners in the stores to move into the data warehouse (a 250-terabyte analytics database that is supposed to be the second largest private database in the world). Up-to-date information about supply and demand helped Wal-Mart reduce both stockouts and overstocking and rapidly reoptimize its merchandising mix; suppliers benefited, in addition, from being able to schedule manufacturing more efficiently and from economies associated with increased throughput.

The Retail Link innovation was clearly related to Wal-Mart's earlier – and early – introduction of electronic data interchange with suppliers, its logistical capabilities which let it contemplate streamlining the supply chain in ways that other retailers apparently could not, and its enormous scale, which generated substantial real and pecuniary economies. (Not unrelatedly, as of 2002, Wal-Mart remained the only source of close to real-time retail data for a large community of suppliers.) Given all these links with Wal-Mart's history, organization and strategy, as well as the potential heterogeneity of combinations implied by the many design choices that had to be made in putting Retail Link together, it is hard to treat it – or more broadly, Wal-Mart's largely home grown, centrally administered, enterprise-wide suite of integrated applications – as generic resources into which it simply poured money. And it is easy to cite many other examples of successful companies that, like Wal-Mart, go to great lengths to ensure that their IT investments are integral to their competitive strategies.

What the arithmetic of capital investment, analogies with other technological revolutions in modern times and examples such as Wal-Mart suggest is that although the risks associated with investments in IT are large, so are the potential rewards. In fact, one could even argue that the aftermath of a stockmarket crash would be a particularly good time to

rethink the real opportunities afforded by a technology such as IT instead
of giving up on them in knee-jerk fashion. That is precisely what this book
by Ricart-Costa, Subirana and Valor-Sabatier attempts to do. Instead of
painting with a very broad brush and asking the overly aggregated question
of whether IT will or will not add value, it recognizes that there will be variations across domains in this regard (just as there were in the case of electricity, for example, in terms of the quick development of opportunities in
lighting at the end of the nineteenth century versus the opportunities that
are still emergent in using electricity to power automobiles).

More specifically, this book presents frameworks for analyzing where
value will reside in information-based industries and the kinds of business
models that are capable of capturing it. It does so by examining the online
value chain from left to right, beginning with Internet content providers and
e-commerce and working through to client software. Each of the ten stages
represents potentially different markets or industries, and in some cases a
stage consists of more than one industry. Each stage is analyzed in terms
that are simple and practical, and that reflect five basic characteristics of
telecoms and information industries: (1) reduction of coordination costs,
(2) reduction of search costs, (3) increasing returns to scale with almost no
variable costs, (4) network externalities, and (5) disappearance of the
reach–richness tradeoff. In parallel, the authors develop a transaction
streams framework throughout the book that provides a contrasting view
centered on the architecture of electronic commerce.

The final, capstone chapter of the book describes how the frameworks
introduced can be used for strategic framing – i.e. in shaping the boundaries
for subsequent strategy analysis. This set of ideas is applicable to many
domains other than the Internet media example that is highlighted in most
of the book. And the treatment is up to date without being technologically
inaccessible For all these reasons, this book can be said to provide a strategic perspective on information technology – and to offer more in the way
of insight than the furious but ultimately unfruitful debate about whether IT
does or doesn't matter.

PANKAJ GHEMAWAT

Pankaj Ghemawat is Jaime and Josefina Chua Tiampo Professor of Business Administration, Head, Strategy Unit, Harvard Business School.

CHAPTER 1

Introduction: Digital Data and the Information Revolution

1.1 Dot.com Realities

Only someone recently roused from a 25-year sleep can be unaware of the ways that information and communication technologies (ICTs) have reshaped our daily routines. To be sure, in the early 1980s there were telephones, television, newspapers, facsimile machines and even computers capable of calculating pi to hundreds of digits past the decimal point. But local and wide area networks did not exist, there were no online research libraries like Nexus or Lexus from which you could retrieve a magazine article or a law opinion, dot.coms had yet to arrive (and depart), and a global computer network called the World Wide Web hadn't been devised – or even considered possible. One could not buy or sell anything from anywhere anytime.

However, by 2004, the business outcome of technology, after a few years of euphoria, has deceived many. The dot.com crash has represented the demise of technology-related business activities. Following it, reports pointing to the demerits of technology abound. The telecoms sector lost $900 trillion. According to a report in December 2001, 70 per cent of IT and e-commerce projects either fail or are completed over budget with less functionality than planned.[1] Spanish Telefónica froze its UMTS[2] initiative with 4.8 billion Euros provisions in July 2002. Controversial technology-bashing is not restricted to its immediate sectors, either. IT now accounts for 50 per cent of all business equipment spending.[3] According to a *Guardian* report, ICTs contributed about 0.7 of a percentage point of the 1 percentage point acceleration in US productivity growth between 1990–5 and 1995–2000.[4] Not all investments in technology seem to be net positive: 1 billion was wasted on failed IT projects in the UK alone.[5] The UK computer giant ICL confirmed that a multi-million pound crew-scheduling system it was developing for British Airways had to be junked after a two-year overrun.

Behind this business outcome are managers at all levels, investors, consultants market research companies and governments. None of them was able to stop the 3G fall out that many felt marked the start of the tech collapse subsequent to the year 2000 Internet stock bubble.

Today, nonetheless, the amount of information – books, movies, photographs, sports scores, songs, games, lists of autos for sale, encyclopedias, hotel rooms availability, video chips, stock market transactions – that can be delivered into (or out of) a business or home by computers and/or other telecommunications devices is virtually immeasurable and would have been unthinkable just five years ago. Reality, again, has surpassed fiction. And the quantity of information being added to electronic networks is estimated to have been increasing by 100 per cent on average every year since 1996.

Statistics usually omit activity that takes place on intranets, but these, too, deliver torrents of information, especially in the business-to-business (B2B) sphere. An intranet is an organization's own network designed for internal sharing, but parts of it can be accessed by outside parties (usually legitimate, occasionally unsolicited hackers) to facilitate a B2B transaction. Ford Motors might be granted access to Firestone Tire's intranet to examine that latter's models, specs, availability and prices, and/or to execute transactions. The network would then be termed an *extranet* – part of a larger network that is limited in access to a pre-agreed set of people. Say you are a small parts manufacturer in Taiwan and you have never sold a thing to Ford but want to do business with them. Ford will go through an accreditation process on you, then grant you access to their extranet so you can start looking at requests for parts makers that Ford has posted there.

This torrent of information and the multiplicity of channels it can take was let loose many years after scientists had figured out how to break down the data that comprise information into electronic signals, compress those signals into manageable portions, send them along cables or over waves, then, at their destination, reconstitute them into information. But what follows is not intended to be a treatise on computer and telecommunication technologies. It's an examination of the value added functions, characteristics and strategies for competitive positioning of the kinds of companies that have adapted or plan to adapt such technologies to the dissemination of information.

1.2 Why Does Digital Information Make a Difference in Business?

Analogue and Digital

Suffice it to begin by noting that there are two modes of conducting information electronically: *analogue* and *digital*. The difference is profound. In an analogue process, data are handled through variable physical quantities like length, thickness, voltage or pressure. Analogue devices measure variable conditions like movement, temperature or sound, and convert them into magneto-electronic patterns, as in a standard television receiver.[6] As the word 'analogue' (from the Greek *ana*, 'according to' and *logos*, 'record') implies, the process – making a musical recording, for example – can produce a similar but not an identical copy. In part mechanical, an analogue recording cannot be made without loss of fidelity, even if this is undetectable by the human ear.

Digital (from the Latin *digitus*, 'finger'), on the other hand, is based on an electric circuit that is either switched on or switched off. The combinations of the familiar '0's and '1's – so-called 'bits' – which today distinguish the codified instructions that manipulate a computer are capable of representing the original source with absolute accuracy. When information – sonic, graphic, or textual – is converted into so-called *binary form* (as, for example, on a compact disc, into which every nuance of every tone of every instrument in a 100-piece orchestra is microscopically etched as a pattern of 1s and 0s), it is no longer recognizable information as such. Rather, it's a unique pattern of 'on' or 'off' commands which can be electronically manipulated, stored and, at dazzling speed, ultimately regenerated into humanly comprehensible output. It is not too much to say that that this simple two-letter alphabet can describe and reproduce practically *all* knowledge, and in the process of doing just that has surpassed English as the world's most widely used business language.[7]

One of its outstanding attributes is that digitized data suffers no degradation between iterations: the nth copy of a given piece of information will be exactly the same as the original. Perfection is not achievable through analogue technology – where, for instance, a subtle distortion in a magnetic pattern may be magnified each time the information is reproduced. (Indeed, the fact that a digital signal can be received perfectly, stored perfectly and relayed perfectly by the receiving party to all members of a network such as the Internet has not only confounded companies whose products are copyright-sensitive, but has launched an industry devoted to intellectual property (IP) copy protection technology.)

Business Information

For business, all this translates into three basic cost characteristics that some believe favour digital over analogue information technology. The former is (1) cheaper to *process*, (2) cheaper to *store* and (3) cheaper to *send*. To these, Andrew Grove, co-founder of Intel, would add a fourth: 'Digital information is forever. It doesn't deteriorate.' Thus the safekeeping and ready retrieval of digitized data, even for companies and government agencies to whom long-term record-keeping is of prime importance and used to be expensive to perform, is no longer a significant cost item.

Metadata

Most importantly, pieces of digital information can be tagged. This brings us to the concept of 'metadata': data describing other data – or, put another way, data that makes other data useful. As digital information, the colours in the sky of a movie scene, for instance, can be inexpensively enhanced by a computer – no longer will Stanley Kubrick have to wait on location for three weeks, as he did in filming *Apocalypse Now*, until nature's real tones are just right. Even more cost-advantageously, full-length digital-from-scratch 'films' (which avoid celluloid and 35mm cameras altogether) are being produced and successfully marketed with budgets perhaps 1/50th that of traditional studio productions. Some movies are skipping human actors altogether, populating their scenes with three-dimensional characters rendered by digitized images.

Computing Power

Some might argue that the Information Revolution is not as vigorous or far-reaching as the Industrial Revolution was in the eighteenth century. Following the invention of the steam engine in the late eighteenth century and its linkage to the cotton gin, the first Revolution rapidly became a huge commercial and social force. But consider the rate at which the Information Revolution has unfolded since the arrival in 1946 of its own innovative 'engine' – a machine called ENIAC (Electronic Numerical Integrator And Calculator). The 18,000 vacuum tubes and 1,500 electro-mechanical relays in what is generally recognized as the world's first computer could perform about 5,000 additions, 33 divisions or 333 multiplications per second. Only 57 years later, a far smaller instrument is capable in one second of performing of 12,300,000,000,000 operations (executions of '0' or '1' transformation commands). Even an ordinary personal computer (PC), the then-novelty gadget introduced in the late 1970s in a popular science magazine as a do-it-yourself project, has become some 176,000 times more powerful than ENIAC, and is doubling in power roughly once every year – a steep trend that some observers predict will continue at least through 2016.

Today's simple desktop box already contains more computing power (the ability to handle large projects) than the IBM mainframe that NASA used to land the Apollo 11 crew on the moon. Computing power, along with storage capacity and bandwidth, are the forces that lie at the centre of the Information Revolution.

1.3 The Online Value Chain: Creating and Appropriating Value in Networked Environments

New Information-Based Relationships

The metrics of business value can – and often do – change rapidly in the expansive and often fuzzy arena of the Internet. A prime example is the record-breaking merger between the world's leading independent service provider (ISP), America Online (AOL) and Time Warner, the world's leading media conglomerate (music, publishing, news, entertainment, cable, movie studios, network services) in 2000 (Exhibit 1.1).

Exhibit 1.1 AOL–Time Warner

Had the merger taken place in the early 1990s, Time Warner would have been in the driver's seat. But at the close of the century their perceived respective values (as measured by market capitalization) had plunged. In 1993, after one year as a public company, America Online (AOL) had generated $4.3 million in net income from $31.6 million in annual revenues and had a market capitalization of only $168 million. Time Warner brought in a net income of $86 million on $13.1 billion in revenues: its market capitalization for 1993 was nearly $11 billion, some 60 times greater than AOL's.

In 1999, boasting some 30 million subscribers around the globe,[1] AOL reported a profit of $1 billion on revenues of $5.7 billion, and the company's market value had jumped to $128.5 billion. In 2003, at 17 per cent it still commanded a market share three times larger than its next competitor. In 1999, Time Warner generated a net income of $1.3 billion on revenues of $23.5 billion still 4 times better than AOL, but the stock market dictated a capitalization of a mere $93 billion barely two-thirds of AOL's. AOL assumed the leadership role in creating and running the enterprise now called AOL–Time Warner. Most observers agree that the merger generated value for AOL customers, but it is more difficult to agree whether Time Warner is actually worth the $106 billion in stock and debt that AOL paid for it, and whether the amalgamated company's Internet media business can capture and hold enough of that added value to make the merger advantageous to stockholders despite the cost.

It would have been difficult to convince observers that the economic future of a corporation which in six years had grown to become the largest IAP in the world was, in a word, bleak. AOL founder Steve Case reached very much that conclusion, however. To sustain AOL's preeminence, continuing to be merely an independent access provider wouldn't do. Content – not just stock quotes and buddy lists, but content that had proprietary value – was essential not only for growth, but for survival (especially in the light of competition from Microsoft's content-delivering high-speed X-Box console). Clients who dialled into the Internet through AOL would be attracted and captured by kind of a pay-per-view menu of movies, video, magazines, music and games that would be delivered through broadband technology. Otherwise the company would be like every other antiquated IAP that gets a phone call and connects the dialler to the Internet, with hardly a concern for the caller herself. So AOL bought Time Warner.

An important component of the AOL–Time Warner combined business model is the independent access provider (IAP). If you're the largest IAP in the world and you're vastly overvalued by the stock market and you believe that content will be essential for you to survive – and proprietary content at that – what do you do? You buy Time Warner. Because, otherwise, you're dead. AOL–Time Warner combines a pure portal with basic Internet access, and that former IAP now can guarantee maximum quality. AOL–Time Warner is its own host, its own backbones, its own IAP, its own portal and its own content.

There are synergies to this model. Separate the AOL portal from the AOL IAP. When you dial AOL, you are funnelled directly into its content in a material way, instead of having to pass through many different intermediaries; further, they add revenues from subscriptions and/or a percentage of phone charges to advertising and tenant-placement revenues. AOL–Time Warner has so far fared better than Yahoo!, maintaining earnings and (relatively) stock valuation.

So uncertain were prospects for success of the pure and IAP models that the portal shakeout began even among well-capitalized companies. Walt Disney, although it was drawing over 13 million separate users a month, closed its flagship portal Go.com after having invested several billion dollars in an Internet division. The top 10 per cent of the remaining portals now derive 71 per cent of the revenues. Some believe that clearly, there is little revenue left over for a third portal model, whatever form it might take.

But AOL–Time Warner competes essentially in most areas of the value chain. A key objective of this book is to provide a framework that helps the reader frame the business challenges facing companies competing in the information industries so that strategic thinking can then be done.

Note:
1. Current ISP statistics can be found at http://isp-planet.com/research/rankings/ usa.html.

The AOL–Time Warner venture is just one example of the rapidly changing value propositions that are being driven by advances in ICTs, exercised mainly on the Internet.[8] The result is the emergence of an interconnected, information-based economy that has created new relationships between sellers of products or services and the customers shopping for and buying them. In this book we will analyse the industries comprising these relationships using a framework which we call 'the online value chain' (see Figure 1.3, p. 11). We shall examine the components of this chain by hypothetically having the client (or user) access a proprietary-content provider. Her electronic commands via keyboard or touch-pad (or possibly speech) will be dealt with by intermediaries along the chain, whose involvement is necessary to fulfil the request, and who have the capability of returning the value that the user is willing to pay for – that is, gaining the desired digitized data in useful and appealing form.

Connecting to the Network

To do so, of course the client must have some kind of device to connect to the network. However, no longer does it have to be the traditional PC; message-receiving mobile phones, satellite-connected personal digital assistants (PDAs), high-speed game-playing equipment and other media are being adapted to Web interchange. In the value chain shown in Figure 1.3, such devices consist of three basic elements: (1) the 'nuts-and-bolts' hardware itself, (2) its operating system and (3) its Web browser; in some instances there is also dedicated special-purpose software that, for example, does the dialling or downloads the data.

Linking this device to the networked universe is the local loop, which connects the user with the Internet Access Provider (IAP). By and large, the local loop today is composed of a traditional twisted-pair of copper wires belonging to the local phone company ('POTS', in Web-speak for 'Plain Old Telephone Service'). Both traditional modem connections and digital links, including ADSL (Asymmetric Digital Subscriber Line – not truly a line but an advanced modem that speeds up reception from the line), travel on these wires. Faster and therefore increasingly preferred alternatives to the twisted-copper 'highway' are cable television, fixed wireless (radio frequency and optical) and mobile wireless. IAPs take the signal from the user's computer and route it to the Web via an access node. The user can request the information she wants by utilizing a browser such as Netscape or Explorer; by finding it via a portal, which is a specifically focused site (such as Yahoo!) that consolidates third-party content and helps surfers locate what they're looking for; or, with progressively more ease, by going directly to the proprietary content provider itself.

1.4 A Simple Transaction Stream[9,10]

The basic framework for most of the book is a linear one. As we shall see this linear framework is very useful in understanding the different stages of the value chain. However, it does not capture some of the non-linear network activity that is prevalent on the Internet and other networks. To understand this non-linear activity we will be using what we call a 'transaction streams' framework. We will give a brief introduction here and expand it throughout the book, mainly in Chapter 8.

Internet Architecture

One alternative way to understand the business activity that is created by a click on a page is by looking at the architecture that enables such transaction. At a very simple level of abstraction, the Internet can be thought of as a cloud on which you can connect any pair of computers so that they can talk to each other (see Figure 1.1). For example, one can connect E-Jo and

The New York Times (*NYT*) and they can talk to each other – i.e. E-Jo types the NYT URL on the computer and, through the Internet, the computer sends a request to the *NYT*. The *NYT* guests the request and then it sends back its home page whose destination is E-Jo's brain. E-Jo may then click on a part of the home page and the dialogue continues.

Modem, ISP and Hosting Computer

At a finer level of abstraction, E-Jo's computer is connected to the Internet via a modem, through the local loop and the ISP (Internet Service Provider) (Figure 1.2). The ISP is then connected with the Internet which can be described as a set of networks that are interconnected to each other. Finally, there is a hosting computer that contains the *NYT*. The *NYT* updates the newspaper by accessing the hosting service and changing the content present there.

1.5 Structure of the Book

The following chapters focus on where network value lies and what kinds of business models are capable of capturing it along the way. Some models have already been superannuated (though they may not yet realize it); no doubt others have yet to be devised. To identify and target the nature of this value, we analyse competitive factors, strategic approaches and trends within each stage of the network, as well as those more broadly affecting the network as a whole. *Business value*, as we define it, refers not only to creating value for the user, but also being able to appropriate part of that value by

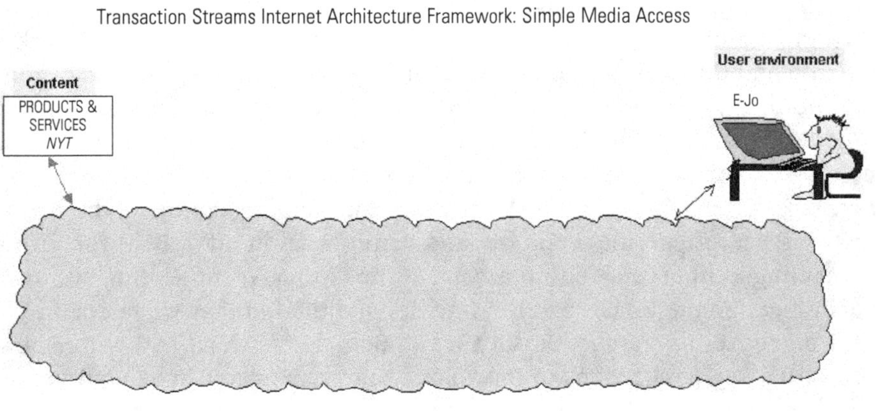

Transaction Streams Internet Architecture Framework: Simple Media Access

Figure 1.1 Transaction Stream I

Transaction Streams Internet Architecture Framework: Simple Media Access

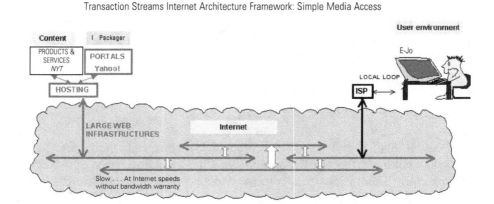

Figure 1.2   Transaction Stream II

generating revenues that exceed the costs of doing business. Of course these profits need not be immediate, but in any model they must be achievable within a survivable time frame, as many now-defunct 'dot.coms' painfully learned in the late 1990s. While we will identify the factors that enable firms not only to capture but to sustain value, we have to caution that in the volatile Internet environment defining factors can and do change.

We first examine the online value chain, beginning with (1) Internet content providers and e-commerce and working through to (10) client software. Each of the 10 stages represents potentially different markets or industries, and in some cases a stage consists of more than one industry. Analysis of each stage begins with a description of the products or companies that compete in that stage, followed by an analysis of key competitive factors, strategic approaches and trends. Because of the value network's breadth and the intricacy of the relationships within it, the analysis cannot go into detail about any one company or market; instead it focuses on the broader issues within each stage and, by extension, throughout the network. The goal is to provide a useful representation of the *online value network environment* (Figure 1.3) and further understanding of the major factors that affect the creation and appropriation of value for firms competing within that environment. After we have done this we will generalize the model beyond the Internet and the Internet value chain.

While covering transaction streams in more depth, Chapter 9 will review one of the key applications of the value chain framework, strategic framing, i.e. shaping the boundaries upon which subsequent analysis can proceed.

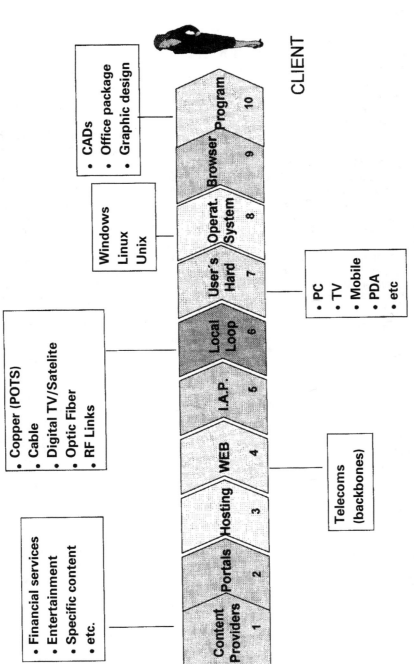

Figure 1.3 The Online Value Network

CLIENT

- CADs
- Office package
- Graphic design

Windows
Linux
Unix

- PC
- TV
- Mobile
- PDA
- etc

- Copper (POTS)
- Cable
- Digital TV/Satelite
- Optic Fiber
- RF Links

Telecoms
(backbones)

- Financial services
- Entertainment
- Specific content
- etc.

Content Providers 1
Portals 2
Hosting 3
WEB 4
I.A.P. 5
Local Loop 6
User's Hard 7
Operat. System 8
Browser 9
Program 10

Notes

1. *Standish Group Report*, December 2001.
2. Universal Mobile Telecommunication Service.
3. Bain & Co., *Business Week*, September 2001.
4. http://www.eco.rug.n./ggdo/pub/gd60.pdf.
5. *Guardian*, July 5 2001.
6. However, in late 1990s, after some 50 years as an analogue entity, the TV industry began turning to digital technology.
7. J. Enriquez, *As the Future Catches You: How Genomics & Other Forces Are Changing Your Life, Work, Health & Wealth*, New York, Crown Business, October 16 2000.
8. The 'Internet' is defined generally as a global network connecting millions of computers. Various components of a network are required for the provision of an expected result. These include interconnected backbones that have international reach, subnetworks and thousands of private and institutional networks connecting various organizational servers that host the information of interest. 'World Wide Web' (also WWW or Web) is a software application that runs on the Internet and enables users to find and store information.
9. For a full comprehension of transaction streams see also: 'Transaction Streams and Value Added Sustainable Business Models on the Internet', in M. A. Hitt, J. E. Ricart and R. D. Dixon (eds), *New Managerial Mindsets: Organizational and Strategy Implementation*, New York, John Wiley, 1998, pp. 129–48.
10. B. Subirana and P. Caravajal, 'Transaction Streams: Theory and Examples Related to Confidence in Internet-based Electronic Commerce', *Journal of Information Technology*, 15(1), March 2000, pp. 5–14.

CHAPTER 2

Client Software

The final three stages in the online value chain in Figure 1.3 (p. 11) – (8) operating systems for user hardware, (9) Internet browsers and (10) software applications – can be reviewed as one entity in that, each stage being composed of companies who develop software applications for end-user hardware devices, they interface with our client directly. These companies include large technology and software entities such as Microsoft and Sun Microsystems, telecoms such as AT&T and Nippon Telephone, specialized software companies such as Intuit and Electronic Arts, Internet-oriented companies such as AOL–Time Warner (with its Netscape browser), as well as many smaller niche competitors.

2.1 Competitive Factors: The Nature of Information-Based Businesses[1]

Software is essentially an *information-based product*; however, it is not the only one. A book, for instance, is a very traditional information-based product, as it is a song or some specialized knowledge. Information-based products give raise to information-based businesses or information industries. As such businesses are very relevant for our purposes in this book, and all of them share on some common characteristics in terms of their competitive factors, we will first study the general competitive factors of information-based business.

Information and Information Goods

We use 'information' in a very general sense to refer to anything that can be digitalized. *Information goods* then include items such as soccer scores, movies or software. There are three fundamental elements that characterize information-based businesses to a greater or lesser extent: (1) the nature of their *cost structure*; (2) the ability to lock in customers by making it too expensive to switch; (3) what we call 'positive feedback'.

Cost Structure

Fixed and Marginal Costs

Information-based products and services are characterized by initially high fixed costs (usually not recoverable) and very low (sometimes practically non-existent) marginal costs from then on. For example, it may cost more than $500 million to develop an operating system (OS), or even more (the Windows XP OS is said to have cost up to $1,000 million), but after that it costs essentially nothing to reproduce it and deliver it to an end-user. AOL can deliver the newest version of its highly intricate service to millions of users, who simply download it: no box, no disk, no shrink-wrap, not even any bookkeeping. (AOL in fact exploits its customers, who willingly act as beta-testers of each upgrade for free.) Much the same holds for most texts, recorded music, films, illustrations, and so on. (Also, these types of artistic and entertainment endeavours don't suffer the added cost of having to support their products.) Online brokerage, it should be mentioned, has exceptionally high fixed costs; they must build an intricate infrastructure that is failsafe, develop the brand, spend to attract fickle clients, and technically maintain their systems; after that, the product is similarly without cost.

Competition and Market Pressures

Any information-based business shares these characteristics: unusually high initial fixed costs, followed by a life-of-the-product period of virtually no cost. This truly constitutes a new economy, if for no other reason than that the marginal cost (refresh your microeconomics!) becomes less important, because the relative effect of the fixed cost is more important. But such a cost structure has some negative consequences as well, because in any industry with low marginal costs and high fixed costs, when there is a lot of competition and market pressure from customers: each market player tends to move toward *commoditization* – selling as many products as they can as cheaply as they can (or in many cases, even cheaper than it cost). This of course provokes price competition, and if everyone allows a product to be influenced only by market price, they're in a dangerous business, because intense price competition destroys all participants.

Monopoly Advantages

But if there's little competition, the opposite occurs: you tend to become a monopoly. If you're the only one, unit costs go down as volume increases – and that's a great business to be in. In this networked economy, one must be aware of cost structure, because it's a different economy when the marginal cost is more important and the relative effect of fixed cost is less important. Whoever tries to enter your market will face discouragingly high fixed costs for their first unit, so that, in already being past that you have a big and usually unbeatable advantage.

The Market vs. Cost Dilemma

Understanding such networked-economy elements is important in applying the technology correctly within a business. One can see the market vs. cost dilemma at work in the proposed 2001 Hewlett-Packard–Compaq merger between two cost-cutting models. The merger did little to alleviate unit-cost problems for either, which (had the merger been voted through) would have left only Dell Computer, with its price-sensitive direct-marketing model and customized ordering provisions, alive in the market. Conclusion: strategic action must be directed toward reducing any tendency to 'standardize' – and thence to commoditize – a product or service, as we demonstrate later.

The Lock-In

Switching Deterrents

The second important element that characterizes corporations in information-based industries is a company's ability to establish a deterrent to customers considering switching to some other company's product line. We call a well-managed, strategic retention of an installed base a 'lock-in'. Information-industry factors that contribute to making your customers think twice about leaving the fold include (1) pre-existing contractual commitments, (2) durable-goods ownership (e.g. of dedicated hardware), (3) specialized assets (for example, proprietary software suitable only for a particular protocol), (4) brand-specific training (having mastered a steep learning curve), (5) databases usable only on proprietary software, (6) access to a company's specialized suppliers, (7) search costs for alternative products, (8) enticement packages that promote loyalty (such as building in discounts for future software upgrades) and (9) capitalizing on the predilection of corporations' traditionally conservative IT purchasing agents to invest only in proven entities.

It's obvious that a customer base such as Netscape's is a meaningful if intangible corporate asset. (Netscape was first into the browser market, and management knew that Microsoft would react as soon as it could load its guns. So in a surprise first strike, Netscape gave away its product free to expand its user base as fast as possible.) But to turn that asset into a powerful competitive weapon, you have to *manage that list*. And that's easier said than done.

Retaining the Installed Base

This brings us to another aspect of the networked economy. Before, when you sold a good you usually calculated the cost, calculated the margin and tried to make a profit on every item you sold. But now you try to make money out of the whole system. Maybe you have to attract customers by essentially paying them to enlist in your system (*à la* Netscape), so that tomorrow you'll be able to earn money from the installed base they constitute. But that's a more difficult flow of capital to manage than in the old economy, and many a now-defunct company has miscalculated, receiving insufficient revenues to recover their up-front investment. The catastrophe most likely to strike is that before you can recover this money, the technology changes and you end up trying to sell yesterday's newspaper. In highly competitive fields – which most are these days – there's little to prevent members of your installed base from jumping ship *en masse*. Even at some cost to them, your clients cannot afford to fall behind.

Positive Feedback

Network Externalities

The third component that characterizes information-based industries is positive feedback, a market state which facilitates the existence of large user bases, or what we call *networks*. Indeed, it is this phenomenon which defines ICTs' influence on the new economy: far more descriptive than the popular term 'new' economy is 'networked' economy. If you own the sole telephone in the world, its only worth is to your ego. But if two other people have one as well, now we can communicate and everybody's phone gains real value. As more phones are added to the network, a very interesting effect takes place. Any new telephone is obviously valuable to the new connected member, but more important, the new telephone raises the value of the whole network and as a consequence the value of each member in the network. This effect is known as the *network effect* or *network externality* because any new member to the network creates an externality to the whole network.

Any real network is subject to this type of externality – think of networks of telephones, faxes or railway trains. In fact the network effect is so important that regulation in these industries includes interconexion rules to avoid the creation of proprietary networks, or the creation of a unique global network. However, network effects are particularly prevalent in information-based businesses. We may not have a physical network, but we have a virtual network usually through the interaction with complementary products or services. The best example are OSs such as Windows. As the number of users for such an OS increases, developers have more interest in developing end-user software that operates on such an OS or more periphericals are produced which are compatible with this OS. Additional value is thus created for the initial OS user.

Economies of Scale

Positive feedback may also be a consequence of economies of scale in supply (which lead to cost reductions for large companies) and economies of scale in demand (a network effect that creates buyer economies). The customer then benefits from having access to increasingly more users of the same product or service. Obviously, the payoff of investing in a telephone increases as the quantity of connected telephones increases and becomes a network. This is the measurable change in benefit to, say, one telephone owner as increasingly more telephone owners join the network.

The positive feedback associated with economies of scale is specially important in information-based businesses because by their nature they give rise to increasing returns to scale. So, while in our physical economy most processes eventually reach their minimum efficient scale and decreasing returns to scale set in beyond this point, in the digital industry there are no limits to scale and therefore we have perpetually increasing returns to scale. The positive feedback of networks that generate economies of scale can be stated in real time, like telephones; or it can be measured in terms of a secondary effect, such as the number of Macintosh users (because the dimensions of a given OS's installed base will be a deciding factor for software companies when developing applications).

eBay

A slightly different dynamic is shown by the case of eBay, whose Internet auction business matches highest-bid buyers with sellers of items and charges a fee at both ends of the transaction. The fact that there is a large omnipresent network of sellers (as opposed to a chance network of buyers) would seem to make those sellers unhappy, since in most businesses sellers don't want yet more sellers to compete against. But the fact that there are increasingly more buyers attracted by the quantity of sellers who turn up is what, in turn, attracts sellers in a self-perpetuating spiral. Seller expansion feels the same as positive feedback, and one can obviously draw the same conclusions. Indeed, in offering an item, a seller would likely choose eBay over, say, Yahoo!, despite the latter's having fewer sellers. There is perhaps no better illustration of the importance of monitoring and nurturing an installed base: eBay's e-business has been able to fend off such would-be contenders as the venerated Sotheby's.

The Market Standard

The company that is first to build a network can take advantage of positive-feedback effects and, concomitantly, IT purchasing agents' tendency to favour single suppliers, which underlie markets there the winners can take all. During the ICT-formative mid-1980s, for instance, this was true for IBM brands in both hardware and software. The usual progression is that after an initial period of competition, one product emerges as the standard and its supplier makes off with the lion's share of clients. From that point, the big get bigger, and the small go elsewhere (often, having exhausted their cash, into bankruptcy court.) Microsoft, for example, got launched toward supreme market domination when, in 1983, its MS-DOS personal computer operating system happened to be chosen by IBM over Digital Research's equally workable CP/M system. Nor is it always superior technology that defines a market standard. Sony's extended campaign to compile a large network in the home video market collapsed when the market rejected its technically superior Betamax in favor of the cheaper VHS.[2]

2.2 Strategic Approaches

The three forces identified in section 2.1 have some common elements which are key to strategies in the software industry. When barriers to entry are relatively low and rivalry is intense among competitors, there is a strong movement towards commoditization of information-based products. When barriers to entry are high, these forces interact to create powerful monopolies that are very difficult to overcome. As can be seen in Figure 2.1, the three strategic factors reinforce themselves, providing a tremendous first-mover advantage to companies able to activate this combined positive feedback loop.

Strategic thinking in these businesses has to appreciate the interaction of these forces that may evolve the business toward monopoly rents or towards intense price competition that lowers market price towards marginal cost with high chances of never recovering the high investment and fixed costs involved. The move in one or the other direction may depend on the existing barriers to entry as well as the speed with which the business can move forward and gain share (and size) very fast.

Three strategies are fundamental in these businesses, with different implications for OSs, browsers or software applications for end-users: (1) setting proprietary standards; (2) managing the lock-in process; (3) avoiding commoditization.

Setting Proprietary Standards

Positive feedback, as we have seen, is the dynamic process by which the strong get stronger. A company able to ignite a positive feedback process will try to manage customers' expectations to get fast subscription to its products and services knowing that the positive feedback will work to the

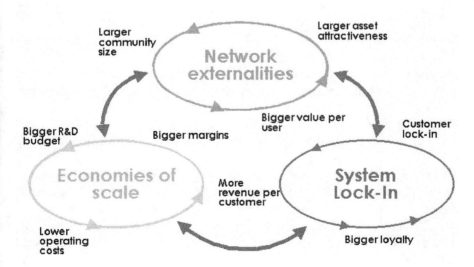

Figure 2.1 A trend towards monopoly power

advantage of large networks and against small networks. Eventually, a piece of software will transform itself into the established standard in the industry with the corresponding system lock-in position for the company.

Developing and Setting a Standard

Microsoft Windows, is the operating systems for PCs, the best example of a proprietary standard established in an industry. In the early 1980s different OSs were competing in the market place. The adoption by IBM in 1981 of MS-DOS as the operating systems for IBM PCs was a fundamental step for Microsoft. IBM designed an open PC, allowing a whole industry of clones to develop. All these elements rapidly increased the MS-DOS installed base, transforming this OS into the industry standard. Only Apple Macintosh was able to survive as an alternative because it was such a superior product in relation to MS-DOS. However, Microsoft successfully managed the transition of MS-DOS to a compatible Windows system able to match the main characteristics of the Macintosh OS, establishing a very powerful proprietary standard in a growing industry.

Setting a proprietary standard is a fundamental strategic approach, especially in OSs. Your ability to successfully wage a standards war depends on your ownership of some key assets, the more important of which is the installed customer base. However, four other key assets are necessary to defend your position:

(1) *Intellectual property rights* (IPRs) as patents and copyrights are strong deterrents to imitation
(2) Ability to *innovate* to avoid a substitutive superior standard to establish itself
(3) *Reputation, brand name and distribution capabilities* are fundamental to reach a broad base of new customers, as well as a lock-in instrument for the installed base
(4) Strength in *complements* can reinforce your position – as the Office software package helps to reinforce the Windows dominance.

To develop and set a standard, it is necessary to act fast to pre-empt competitors and ignite the positive feedback to work to your advantage. It is a risky business that requires high commitment, ample resources and tolerance of risk. Netscape tried to establish its browser (Navigator) as the standard by giving it away with the expectation of reaching high benefits once the standard was accepted. However, Microsoft did not want to risk its current dominant position in the PC industry by allowing a competitor to control a fundamental standard in the Internet. Microsoft developed and introduced the Explorer. Thanks to different bundling strategies with Windows, Microsoft Explorer was able to overcome the Navigator as the browser of preference. As a consequence Netscape was not able to obtain the desired benefits after the effort it made to introduce its browser (see Exhibit 2.1).

Expectations

Expectations are a key factor in consumer decisions about the purchase of any new technology product. As a consequence, to set a standard, one has to learn how to manage such expectations effectively. Microsoft again is king in the management of expectation because most consumers grant them the victory even before the battle starts.

Defending a Position

Once the standard is established there is still a great deal of strategy involved in defending your current position. Even with Windows' dominance, it is still vulnerable to some actions of competitors. We have already discussed how the development of Windows tried diminish any possible advantage associated with the Macintosh OS. Windows, however, had other attacks to meet. Unix is an alternative OS that has been increasing its share, especially among servers. Once Netscape developed the Navigator, Microsoft realized that the browser provided an interface compatible to any of the existing OSs and applications could be developed that would break the system lock-in associated with Windows. Microsoft fought very hard, leveraging on Windows' dominant position with the Explorer.

Exhibit 2.1 The Browser War

Introduction

Browsers are software applications that enable information exchange and commercial transactions over the Internet. Internet Explorer (Microsoft) and Netscape Navigator (Netscape Communications, acquired by AOL in March 1999) are the two most widely used.

The development in 1993 of the graphical browser Mosaic by Marc Andreessen and his team at the National Centre for Supercomputing Applications (NCSA) gave the Internet protocol its big boost.

Andreessen moved to become the brains behind Netscape Communications, which produced what was the most successful graphical type of browser and server (Netscape Navigator) until Microsoft joined the battle and developed its Microsoft Internet Explorer in 1995.

The History of Netscape Navigator

As a student at the University of Illinois-Champaign, Marc Andreessen helped develop a prototype for Mosaic, an easy-to-use browser for UNIX-based computer networks. Mosaic helped spark an explosion in Internet use and Web site development. In 1994 the company became Netscape Communications. Navigator, freely distributed, became the dominant Web browser. Netscape made money by selling upgraded versions of Navigator; it also sold Web server software to corporations.

Even though it had never made a profit, in 1995 Netscape went public in one of the hottest IPOs[3] of the decade: its first-day market capitalization was $2.2 billion. When Netscape's stock plummetted from its high 1995 values, CEO James Barksdale cut his pay from $100,000 to $1 (but not before cashing in $100 million of his own stock).

Since then Netscape has tried to stay focused on corporate markets, but the acquisitions that have been made to achieve that, combined with the stiff competition from Microsoft and slower growth, resulted in huge losses in 1997. That year several Baby Bells (Ameritech, Bell Atlantic, BellSouth and SBC Communications) agreed to make Netscape Navigator their primary Internet browser. However, AOL stung Netscape when it chose Internet Explorer. An intense market attack from Microsoft forced Netscape to give away its browser for free rather than selling it, and led to layoffs of more than 12 per cent of its workforce.

In 1998 Netscape teamed with Internet search engine company Excite to create co-branded content and search services on Netscape's Web page in a joint effort to unseat Yahoo! as the first navigation guide. In 1999 Netscape was acquired by AOL and Andreessen was named AOL Chief Technical Officer.

The History of Internet Explorer

Bill Gates and Microsoft were slow to recognize the significance of the Internet. Microsoft's first release of Internet Explorer 2.0 was just a slight modification of Mosaic. Internet Explorer version 3.0 in early 1996 was the first serious threat to Netscape's dominance. The fact that the product was offered free of charge helped fuel demand for the program, which was downloaded 50,000 times a day when it was launched.

With Explorer, Microsoft did not create the best Internet browser on the market; but by putting huge human and financial resources behind its product, as well as sheer marketing muscle, Microsoft emerged as the winner. Today Internet Explorer is the dominant player in the browser market while Netscape has a significantly reduced market share.

Apple rolled out the Safari turbo brower for Macintosh computers in 2003, promising outstanding performance and speed.

Browser market share

There is no agreement about the different browsers' market share. *AdKnowledge* data shows that Netscape lost approximately 30 per cent market share between January 1997 and August 1998, while MDC data suggest a more modest decline.

Until the introduction of Internet Explorer 3.0 in September 1996, Internet Explorer had an essentially flat market share of less than 10 per cent, even though Internet Explorer was included in all versions of Windows 95 and an Internet Explorer icon appeared on the Windows desktop. Between 1998 and 1999 Microsoft became the market leader and Netscape began to lose market share year by year (Figure 2.2).

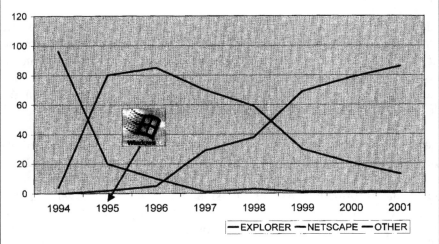

Figure 2.2   Internet browers' market share

Source: AdKnowledge, Dataquest and Websidestory, August 2002.

Sun Microsystems and others pushed very hard on the use of Java as a language to develop applications based on any browser so that it is essentially compatible to any OS. The broad acceptance of Java for the development of applications is a very serious threat to Microsoft's dominant position, especially as the combination of Unix and the Navigator makes Windows unnecessary. Microsoft defended itself by again leveraging its dominant position in Windows and Explorer, making the browser fail with some Java applications. This uncompetitive move was defeated in court by Sun, and Explorer is now Java-compatible. However, Microsoft is now using all its power to develop what it calls the .Net strategy, by developing an alternative C# language to compete with Java.

All this shows that the owner of a proprietary standard still needs to be on guard, trying to see where the new generation of technology is going, and trying to avoid movements in other software stages of the value chain commoditizing our controlled stage in the value chain. We have seen how the fight has moved from the OS, to the browser, to the language for user applications. Microsoft fought on each battleground.

As the best defence is an attack, the owner of a proprietary system must make all efforts to commoditize other stages of the value chain and as well as key complements they may not control, because the control and command of complementary products is fundamental in a system lock-in strategy.

Driving Innovation

Another important element in a defensive strategy is to learn to compete with your own installed base, as a way to keep growing when the information product starts to reach market saturation. The key is to drive innovation even faster by upgrading the products constantly, something that happens in Windows upgrades and new generations, as well as in the development of Microsoft Office, for example.

At the end of November 2002 all industry newspapers echoed with stories on the revenues and margins of Microsoft. In the third quarter of the year, while the Client Division, including Windows, was making $2,480 million on $2,890 million revenues (equivalent to margins over 85 per cent), the Information Worker Division, including Office, was making $1,880 million on $2,390 million revenues and the Server Platforms was making $519 million on $1,520 million revenues. But what caught the attention of the press were the divisions with high losses such as MSN with $97 million on $531 million revenues, Home and Entertainment with $177 million on $505 million revenues, Business Solutions with $68 million on

$107 million revenues, or the CE/Mobility Group with $33 million on $17 million revenues. Even adding up all the losses from all divisions, it is a low figure compared with the high margins of profitable divisions.

The key for us is that Microsoft is willing to compete in all fights that might put its virtual monopoly at risk. Whether it is with its MSN strategy or developing OSs for any other hardware that may potentially substitute the PC – be it the Internet itself, the game consoles, the TV, the mobile phone or the PDA – Microsoft is playing everywhere to defend its standard (see Exhibit 2.2).

Exhibit 2.2 Microsoft

The History of Microsoft

Bill Gates and Paul Allen founded Microsoft in Redmond (Washington) in 1975.

Bill Gates dropped out of Harvard at age 19. Bill and Paul Allen made a version of the programming language BASIC for the first commercial microcomputer, the MITS Altair. Their business idea was to create software systems that let people have personal computers on every desk and in every home. Microsoft grew by modifying BASIC for other computers. Gates moved Microsoft to Seattle in 1979 and began developing software that let others write programs.

The modern PC era dawned in 1980 when IBM chose Microsoft to write the OS for its new machines. Instead of developing a new system, Bill Gates bought an OS in the market from a local programmer in Seattle and then adapted it to the Intel microprocessor, which became the heart of the IBM PC. The Microsoft operating system was called Microsoft Disk Operating System (MS-DOS). In 1984, MS-DOS achieved 85 per cent market share. The company went public in 1986, and Gates became the industry's first billionaire a year later.

In 1990 Microsoft introduced Windows 3.0, which became the dominant operating system for Intel-based PCs. In the mid-1990s Microsoft had three competitors in the operating system market: Apple, with 8.5 per cent market share; IBM OS/2, with around 5 per cent; and different UNIX versions, with 3–5 per cent.

When the Internet began to transform business practices, Gates embraced the medium and Microsoft launched Internet Explorer (an Internet browser) and Microsoft Network (an instant messaging program) in 1995. The release of Windows 98 in June 1998, with Internet Explorer well integrated into the desktop, showed Bill Gates' determination to capitalize on the enormous growth of the Internet. Microsoft's success since the late 1990s has been accompanied by court challenges to its dominance.

In 1997, Sun sued Microsoft for allegedly creating an incompatible version of Java. A federal judge ruled later that year that Microsoft had used its monopoly powers to violate antitrust laws. Microsoft aggressively appealed the decision.

In 1998, the US Justice Department, backed by 18 states, filed antitrust charges against the software giant, claiming that it stifled Internet browser competition and limited consumer choice. In 2001, the settlement left Microsoft intact, but imposed broad restrictions on the company's licensing policies for its operating systems.

Netscape Communications (a division of AOL Time Warner) filed suit in 2002 against Microsoft, seeking damages and injunctions against the company's antitrust actions related to Windows and Internet Explorer. An initial ruling to split Microsoft into two companies was struck down, leading to a tentative settlement between the company and the US Justice Department. Under the terms of the pending settlement (the Judge Colleen Kollar-Kotelly ruling) Microsoft would uniformly license its Windows OSs, cease to offer exclusive contracts with manufacturers, and allow competing software to be included with its OSs.

In 2003, the company announced that it would pay out its first dividend. In this way, Bill Gates responded to the demands of its shareholders, who had been asking to receive cash from the company.

Now Microsoft is worried that the success of the open source movement could hurt its sales and market share. The company has already agreed to open its usually secret source code and has recently disclosed technical information known as application programming interfaces. In order to deal with these problems, Microsoft has recently increased its research and development (R&D) budget in order to improve its software, revenues and operating profit (Tables 2.1, 2.2).

Product Portfolio

While desktop applications and platforms remain the cornerstone of its operations (responsible for about two-thirds of sales), Microsoft has inexorably expanded its product lines, which include video game consoles, enterprise software, computer peripherals, software development tools and Internet access services.

The company has also used acquisitions to expand its enterprise software offerings, which include applications for customer relationship management (CRM) and accounting. Along with rival enterprise software providers such as SAP and PeopleSoft, Microsoft is increasingly targeting small businesses.

Table 2.1 Microsoft Corporation, financial statistics, June 1998–June 2002

All amounts million dollars except per share amounts	Jun 02	Jun 01	Jun 00	Jun 99	Jun 98
Revenue	28.365	25.296	22.956	19.747	14.484
Cost of Goods Sold	4.107	1.919	2.254	1.804	173
Gross Profit	24.258	23.377	20.702	17.943	14.311
Gross Profit Margin (%)	86	92	90	91	99
SG&A Expense	11.264	10.121	8.925	6.890	6.347
Depreciation & Amortization	1.084	1.536	748	1.010	1.024
Operating Income	11.910	11.720	11.029	10.043	6.940
Operating Margin (%)	42	46	48	51	48
Nonoperating Income	−397	−195	3.090	1.688	473
Nonoperating Expenses	0	0	0	0	0
Income Before Taxes	11.513	11.525	14.275	11.891	7.117
Income Taxes	3.684	3.804	4.854	4.106	2.627
Net Income After Taxes	7.829	7.721	9.421	7.785	4.490
Continuing Operations	7.829	7.721	9.421	7.785	4.490
Discontinued Operations	0	0	0	0	0
Total Operations	7.829	7.721	9.421	7.785	4.490
Total Net Income	7.829	7.346	9.421	7.785	4.490
Net Profit Margin (%)	28	29	41	39	31

Source: Hoover

Table 2.2 Microsoft Corporation, revenue and operating results, 2001–2

(million of dollars) Business Units	Revenue 2002	2001	Operating Results 2002	2001
Client ($)	4.751	5.327	3.822	4.447
Server Platform	2.774	3.188	752	1.017
Information Worker	4.134	4.796	3.213	3.762
Business Solutions	147	246	(80)	(161)
MSN	893	1.100	(410)	(254)
CE/Mobility	31	38	(109)	(72)
Home and Entertainment	1.069	1.787	(248)	(525)

Source: Microsoft Corporation, *Quarterly report* (period ended 31 December 2002) sent to SEC.

In Table 2.2:

- **Client**: includes revenue and operating expenses associated with Windows XP Professional and Home, Windows 2000 Professional, Windows NT Workstation, Windows Me, Windows 98 and embedded systems.
- **Server Platform**: consists of server software licences and client access licences (CALs) for Windows Server, SQL Server, Exchange Server, Systems Management Server, Windows Terminal Server and Small Business Server. It also includes BackOffice/Core CALs, developer tools, training, certification, Microsoft Press, Premier product support services and Microsoft consulting services
- **Information Worker**: includes revenue from Microsoft Office, Microsoft Project, Visio, other standalone information worker applications, SharePoint Portal Server CALs and professional product support services.
- **Business Solutions**: include Microsoft Great Plains, bCentral and Navision.
- **MSN**: includes MSN Subscription and MSN Network services.
- **CE/Mobility**: includes Pocket PC, Handheld PC, other Mobility and CE operating systems.
- **Home and Entertainment**: include Xbox video game system, PC games, consumer software and hardware and TV platform.

Managing the Lock-In Process

A related strategy is managing well the lock-in process. Locking-in customers thanks to proprietary standards is a systems lock-in approach. However there other switching costs that one may try to manage or create to lock-in customers. For instance, in the airline industry, under big pressures from price competition after liberalization, airlines developed the frequent flyer programs as a way to try to lock-in profitable customers. Some recommendations for customer lock-in, widely used in the software industry are:

1. Be prepared to invest to *build an installed base* though promotions and by offering up-front discounts. Before you can aim for customer entrenchment, you need customers willing to sample your products so that you can build an installed base. You must look ahead in the lock-in cycle and fight for new customers, cultivating influential buyers that

may recommend your product, and buyers with a high switching cost so that it is more difficult for them to leave you.

2. Design your products and your pricing to get your customers to *invest in your technology*, thereby raising their own switching costs. You need to develop creative ways to increase customer switching cost by making files difficult to translate to alternative technologies, by collecting key data from your client, by personalizing the service, etc.

3. *Maximize the value of your installed base* by selling your customer complementary products and selling access to your installed base. At the end, you must be able to leverage on your installed base by selling upgrades, new products, additional services, etc.

Managing the lock-in process well is very difficult. Netscape tried it with Navigator (see Exhibit 2.1) but Microsoft's aggressive response made it very difficult for the proposed strategy to succeed. Microsoft has leveraged its dominant position in Windows to lock-in MS Office customers by integrating different applications with specific characteristics of the OS. The new antitrust ruling (the Judge Kollar-Kotelly ruling), is trying to moderate this by opening some of the source programs in Windows, making it easier to develop applications running smoothly on Windows.

Avoiding Commoditization

While strategies to set proprietary standards and to manage the lock-in process are sometimes very difficult to apply, all software producers need to fight against the fast commoditization of the product or service being offered. Recall that competition in industries where marginal costs are very low will tend to drive competitive prices close to zero, making the whole industry very unprofitable.

Pricing Strategies

There are three basic pricing strategies that can be used as complementary moves to your other strategies studied above:

• The first one is *personalized pricing*. All efforts to differentiate your product and personalize it are very important. You must know your customer, adapt to her characteristics, and personalize the product and the price as much as possible. In this way, you drive competition towards differentiation and avoid destructive price competition.

- A second common pricing strategy is what economists refer as *third-degree price discrimination*, where instead of personalizing you try to discriminate different segments of clients and apply group pricing.
- A third pricing strategy is *versioning*; this is offering your product in different versions for different market segments. There many different dimensions one can use to version products, from delays, user interfaces, image resolution, speed of operation, format . . . If your market segments in a natural way, use it to design your product line. Otherwise, it may be advisable to choose three versions.

Bundling

When facing high variation in willingness to pay for different products, bundling them may be a good alternative. By decreasing the variation across customers in their willingness to pay, combining complementary products we can increase revenues.

As we saw in Figure 2.1, the three competitive factors of economies of scale, customer lock-in and network externalities, reinforce each other, creating a strong trend towards monopoly power. Therefore, the strategic approach has to be able to battle simultaneously on each of the three fronts. It is this combined management of standards, customer retention and strategic pricing and product design that unlocks the door to a dominant position in information-based products.

2.3 Industry Structure and Trends

The three final stages in the value chain elaborated in this chapter – (8) operating systems, (9) browsers and (10) software applications (see Figure 1.3) – share the same characteristics in terms of competitive factors. We have seen how the development of these stages has been affected by the competitive actions of companies in their fight to set proprietary standards or to defend against the proprietary standards of competitors – an interesting combination of competitive defences towards monopoly dominance, especially of Microsoft, but also to avoid or delay the competitive trend towards commodity status (commoditization).

Paradigm Shifts

The rapid pace of technological change is leading to three possible paradigm shifts: (1) a shift from PC-based software applications to server-based or web-based applications, (2) a shift from proprietary software to open source software and (3) a shift from PCs to alternative Internet devices:

A Shift from PC-Based Software

Windows is the king in the PC environment. Microsoft reinforces its position by controlling most of the PC-based software applications. If these applications move from the PC to the server or the web itself, Microsoft's situation will drastically change. Competitors see clearly that this is one way to overcome Microsoft's dominance and, as a consequence, they have been strongly pushing in this direction. But Microsoft is not an easy competitor and reacted with Explorer and the so-called 'browser war' (Exhibit 2.1). Later on it developed the .Net platform and started another war against the dominance of Java to develop web applications. The battle is still on both in the market place and in the courts with recent rulings in benefit of Sun Microsystems, as we saw above (Exhibit 2.2). The outcome of such battles may be a new paradigm where application software will move from the PC to the web (see Exhibit 2.3).

Exhibit 2.3 J2EE vs. .Net

Web services and Web applications

The Internet is obviously a valuable tool for basic communications and commerce, but it has been underused by enterprises. To leverage the true power and presence of the Web in businesses, a more robust, intelligent and easily integrated architecture is needed. Integrating applications, the Internet and data sources in companies can be complex, expensive, and time-consuming.

The idea of developing a common application interface that enables disparate devices to integrate with applications and data much more easily has become more pervasive with the presence of new mandatory tools for businesses, personal digital assistants (PDAs, like Palm Pilot) and mobile phones.

Web services are business and consumer applications delivered over the Internet that companies can select and combine through almost any device, from personal computers to mobile phones. By using a set of common protocols and standards, these applications permit disparate systems to share and integrate data and services – without requiring human beings to translate the conversation. The result promises to be real-time links among the on-line processes of different companies. These links could shrink corporate IT departments, foster new interactions among businesses and create a more user-friendly World Wide Web for employees.

Integrating Web-based applications into desktop business architecture and applications has been more of a challenge than many companies anticipated, and e-business companies have been focusing on fairly simple applications such as e-mail and Web sites. If Web services reach their full potential, they will change the way many companies do business. Reduced transaction costs will encourage the outsourcing of non-core functions through electronic links with key partners. The likely result will be increased fragmentation of value chains and industries, as well as more narrowly focused companies.

Microsoft and Sun Microsystems are making substantial investments in Web services. In this area, J2EE has been developed by Sun Microsystems while .NET has been developed by Microsoft. Both applications try to integrate enterprise systems by developing Web applications or Web services, moving away from PC-based applications, which have been controlled by Microsoft for the last decade.

Microsoft .NET

In 2001, Microsoft announced an initiative designed to integrate its Internet and Web-based applications and services in order to build agile, integrated and effective businesses. This new business strategy is called .NET, which is a software platform that connects information, people, systems and devices. It provides the framework for an Internet-centric industry standard that will make it easier to tie together applications, information, data sources and services across the Web using XML Web services. .NET is built on XML (Extensible Markup Language), which is a new application industry standard.

What is unique about these services is that .NET-connected applications will automatically find, connect and collaborate with them, making employees productive and informed. For example, .NET can provide the foundation to link customer-facing applications to order and inventory systems to ensure that customers placing orders will receive the goods within the specified time frame.

Among the applications and services currently available or planned for future availability are the following:

- Microsoft Office XP: Microsoft's Office suite of desktop applications (including Word, Excel, Outlook and others) are .NET connected and are easily integrated into other .NET-connected applications and services.
- Microsoft accounting applications are all planned to be .NET compatible.
- Microsoft CRM (customer relationship management): Microsoft's CRM suite of applications will be built on the .NET Framework. .NET compatibility will enable Microsoft CRM to access customer information or leads harvested from a Web site, for example, and pass those leads on to the sales team as part of the Microsoft CRM package functionality.
- Microsoft .NET password: a .NET Web-based service available to consumers and business users that provides authentication and verification services.
- Microsoft MapPoint .NET: MapPoint .NET is an XML Web service that enables a .NET-connected application to integrate dynamic mapping and location services as an integral part of its functionality.

Example of .NET-connected applications: HVAC. Inc.

HVAC, Inc. is a privately owned and operated heating, ventilation and air conditioning contracting company. HVAC is using Microsoft .NET applications for order entry, invoicing, accounts payable and receivable, and inventory management. Microsoft CRM is used by the HVAC marketing department's in-house salespeople and field support (see Figure 2.3).

HVAC, Inc. has engaged a firm in Texas that uses the Internet to search for companies that have applied for new building and construction permits nationwide. This information is sorted by ZIP code, provided as a .NET-connected XML Web service, and sold as leads to a variety of construction and maintenance firms like HVAC, Inc. New leads (Lead Harvesting) go directly to the HVAC, Inc. CRM system, using Microsoft CRM's .NET facilities. The system automatically contacts a credit verification service over the Web to do a credit check for the estimated amount of work. .NET credit verification is their

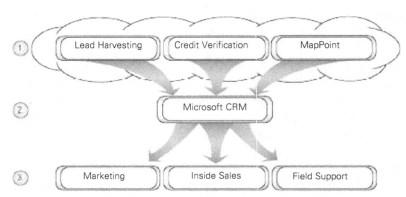

HVAC, Inc. uses .NET services to locate, close, and implement a new customer

Lead management service is located through the Web and provides qualified leads to telemarketing via links for pre-qualification.

Microsoft CRM passes lead on to inside sales group where the customer is qualified, credit vertification is processed in the background, and the order is cleared and entered.

Field support organization is notified of a new customer via Microsoft CRM, and uses MaPoint.NET links to the customer record to identify location and plan the best route for several customer implementations.

Figure 2.3   HVAC, Inc. and Microsoft.Net

Source: Aberdeen Group, 'Microsoft.NET: A foundation for Connected Business', June 2002.

provider in these activities. Once the order has been processed, HVAC, Inc. uses MapPoint .NET to find the location of new customers and to plan the routes and work schedule for the week. The Microsoft CRM system automatically prints a map of the customer's location, a copy of the original sales order and an invoice for the service person to drop off when the system is installed and operational.

Web services built on a .NET infrastructure and using services provided by Microsoft and by Microsoft alliance partners and developers have the potential fundamentally to change the way medium-sized enterprises think about business computing.

Sun Microsystems J2EE

Sun Microsystems, Inc. has been committed to open systems and network computing since it was founded in 1982. Their corporate strategy is to integrate devices: cell phones, desktops and vending machines in order to leverage the opportunities of the Internet.

Java, their most famous product, is the cross-platform architecture for building network-centric applications and integrating these devices. With Java, the industry can take advantage of a technology that is defined in an open and collaborative manner. But Java does not integrate applications and services across the Web using XML Web services like Microsoft.NET. The Java platform provides robust end-to-end solutions for networked applications as well as a trusted standard for embedded applications. It includes three editions: enterprise edition (J2EE) and standard edition (J2SE) and micro edition (J2ME).

The JavaTM 2 Platform, Enterprise Edition (J2EE) defines the standard for developing multitier enterprise applications. J2EE simplifies and integrates enterprise applications by basing them on standardized, modular components, by providing a complete set of services to those components and by handling many details of application behaviour automatically, without complex programming.

Microsoft licensed Java from Sun in 1996 and began shipping it with Windows a year later. But from the start Microsoft began to modify the technology and programs. Claiming that the changes violated its license agreement, Sun sued Microsoft in 1997. That case was settled in 2000, but Sun sued again in March, accusing Microsoft of violating the earlier settlement. Java was not included in Windows XP when that operating system was released in October 2001. But in response to Sun's latest suit, Microsoft again began including its older version with the release of an XP update, called XP Service Pack 1.

On 8 March 2002, Sun Microsystems filed suit against Microsoft again alleging efforts to acquire, maintain and expand a number of monopolies. Sun Microsystems required Microsoft to distribute implementation of the Java software as part of Window XP and Internet Explorer. Recent rulings have benefited Sun Microsystems; a federal judge ordered the software giant to include the latest version of Sun Microsystems' Java in every copy of Windows XP, rather than the 6-year-old, Microsoft-modified version now included.

The dominance of some software applications may also have a fundamental impact on the basic software pieces as the operating system or the browser. For instance, the success of companies such as SAP in ERP (see Exhibit 2.4) can develop systems that totally dominate the OS.

Exhibit 2.4 SAP

ERP

ERP (enterprise resource planning) systems are information systems that use an integrated database to support typical business processes within functional areas and permit consistent information access across areas. These systems provide the integrated view necessary to coordinate purchasing and production scheduling activities.

ERP systems have received much attention for their potential to help companies make more effective decisions. Throughout the 1990s, large industrial companies installed ERP systems that promised huge improvements in efficiency (shorter intervals between orders and payments, lower back-office staff requirements). Chief Information Officers were attracted by the opportunity to replace complex, disparate and obsolescent applications with a single system. ERP systems helped them make better-informed decisions. Encouraged by these possibilities, businesses around the world invested a total of around $300 billion in ERP during the last decade. For many firms, the implementation of ERP systems was traumatic. Following long, painful, and expensive implementation, some companies had problems when they tried to identify any measurable benefits. They also had to introduce a continuous-improvement mindset, which was not popular with employees.

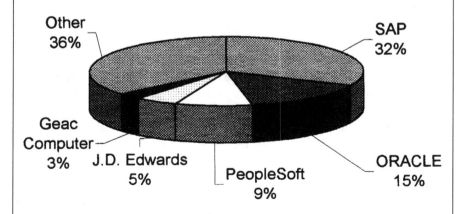

Figure 2.4 Top 5 ERP vendors, 2000

Source: AMR Research 2001 – Total Revenues 2000.

In 2000, the most important companies in the ERP market were SAP, Oracle, PeopleSoft, JD Edwards and Geac Computer (Figure 2.4).

SAP

SAP is one of the most important companies in the ERP global market. More than 17,500 companies use its software. SAP has leveraged its prominent position in the ERP market to expand into related fields, including supply chain management and customer relationship management. It has also developed enterprise portals and online marketplaces in conjunction with partners. In 2000, partly in response to Microsoft's expanding presence in the enterprise software market, SAP released a scaled-down version of its software designed specifically for small businesses. The company also expanded its strategic partnership with Commerce One, an electronic market-place specialist, raising its ownership stake to about 20 per cent.

In 2002, the company reabsorbed its SAP Portals and SAP subsidiaries, integrating their offerings into its mySAP product family.

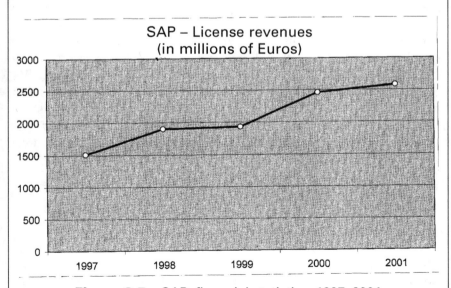

Figure 2.5   SAP, financial statistics, 1997–2001

Source: SAP, 2003.

Now SAP is pursuing a broader vision that encompasses not only all
SAP applications, but the entire world of business technology. It is a
vision of fully integrated solutions that provide the infrastructure
needed to conduct e-business projects.

Despite the economic downturn, sales have continued to grow
substantially, by 17 per cent in 2002. Three of SAP's founders – co-
chairman and co-CEO Hasso Plattner, Dietmar Hopp and Klaus
Tschira – control about 35 per cent of the company (Figure 2.5).

A Shift from Proprietary to Open Source Software
The *open software movement* is a very interesting development in the software industry. There have always been pieces of open software but developments as Linux, a Unix-based operating system, are showing that open
software may be the cornerstone of a new paradigm shift. Linux has been
developed from an initial source with contributions by many independent
software developers and individuals and coordination by a centralized
approval of any new addition. This new organization to develop a very
expensive piece of software has been able to compete against a powerhouse
as Microsoft. Linux (see Exhibit 2.5) is now a serious competitor to
Microsoft Windows, especially on the servers front. People seem to relate
better to the freedom of Linux than to the dominance of Microsoft. And this
is a movement not just on the supply side, but also on the demand side of
the market. Some companies are openly stating they want to reduce their
dependence from Microsoft and are embracing open source software. A
Goldman Sachs Group report in 2002 indicated that 39 of the *Fortune 100*
US companies were using open source software. Even some governments,
for example, the Sweden government, find the dependence of their systems
on one company unbearable and are also actively working in favour of open
software. Microsoft has reacted by opening 97 per cent of its basic sources
– a signal of a new paradigm shift.

Exhibit 2.5 The History of Linux

Open source software

Open source software means that intellectual property rights (IPRs) to
software code are deposited in the public domain. The code can be used
and changed without requiring a user fee, such as the purchase of a
licence. The production model or programming work is accomplished

in a distributed and dispersed Internet community. Therefore, open source software exploits the *distributed intelligence* of participants in online communities and innovation occurs on a more distributed basis.

This model is efficient because it avoids the high fixed costs of a strong IP system and implements design and testing of software modules concurrently. Because open source software works in a distributed environment, it presents an opportunity for developing countries to participate in the innovation process (Figure 2.6).

	Open source	Commercial software	Impact
Development process	Scientific method	Top-down	Open source projects can benefit from thousands of developers.
License terms	Typically free	Pay by the server	The license cost of deploying open source is significantly lower.
Feature selection	The community, led by a maintainer	Marketing and business managers	Portability improves, but at the expense of some desirable features.
Pace of innovation	Always incremental	Often incremental, but sometimes blessedly disruptive	Open source is good for stable versions of known software.
Relationship to standards	Standard by definition	Standards catch on only as a last resort	Open source accelerates the adoption of software standards.

Figure 2.6   The impact of Open Software

Source: Forrester Research, January 2003.

The main disadvantage of open source software is that many people think that it is difficult to use and so they avoid adopting it. Businesses and universities are taking too long to move on to open source software due to their lack of knowledge and the problems they have when they want to install and use it.

The History of Linux

Linux is a computer operating system developed in the 1990s by Linus Torvalds, a computer science student at the University of Helsinki in Finland. Linux was an attempt to build a free PC version of the UNIX operating system, which was developed at Bell Labs in 1969. Torvalds began his work by releasing the first version of Linux in 1991. Instead of securing property rights to his invention, Torvalds posted the code on the Internet with a request to other programmers to help him to upgrade the version into a working system. In this way, it became open source software. The response was overwhelming. What began as a student's project rapidly developed into a non-trivial OS. This was possible because at that time there was a large community of UNIX developers who were disenchanted with the way vendors had taken over UNIX development, they also were unhappy with the growing reliance on Microsoft's proprietary server software.

The Linux development model is built around Torvald's authority, described by some as 'benevolently' exercised. Legally, anyone can build an alternative community to develop other versions of Linux. In practice, the development process is centralized. Linux source code is freely available to everyone. However, its assorted distributions are not free, companies and developers may charge money for it as long as the source code remains available.

Linux is the most rapidly proliferating open source OS in the world and even Microsoft acknowledges it as a threat to its market dominance, especially in the server market because Linux is often considered an excellent low-cost alternative to Windows. Linux may be used for a wide variety of purposes including networking, software development, and as an end-user platform. Linux may still be a mystery to people who just use their PCs for word processing, but Linux is well known among those responsible for overseeing servers. Major computer suppliers like Dell, H-P, IBM, Oracle, Silicon Graphics and SAP (see Exhibit 2.4) are lining up to endorse it (Figure 2.7).

Linux has to make it easy to move from Windows to Linux, just as Microsoft made it easy to move from DOS to Windows. In other words, Linux has to make possible for consumers who are using existing Windows applications (Word and Excel, for instance) to switch to the Linux equivalents.

Several companies want to make the system more widely used. Recently, SCO Group, Red Hat and SuSE have made a business out of tweaking the basic Linux system with their own proprietary software to make Linux easier to install and use.

Red Hat has announced in 2002 that its next version of Linux was to feature an easier interface to make the software friendlier to PC users at home and in schools. IBM and Red Hat also announced a multiyear alliance to market Linux. The two companies have said they will join forces to provide Linux support to IBM enterprise customers around the world. Among other things, Red Hat agreed to deliver support for the Red Hat Linux Advanced Server on IBM's zSeries, iSeries and pSeries of eServers.

In addition, Sun Microsystems plans to offer customers package deals that include Linux, desktop and server hardware and application programs.

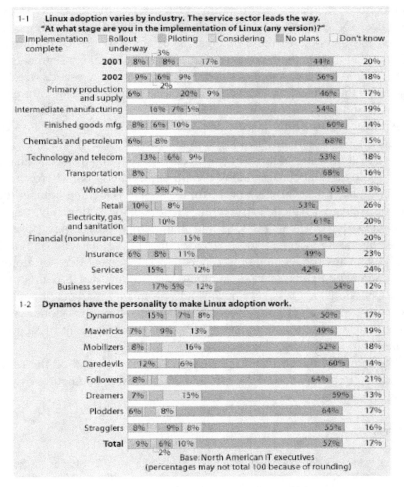

Figure 2.7   The implementation of Linux

Source: Forrester Research, July 2002.

A Shift from PCs to Alternative Internet Devices

The PC is the dominant device to access the Internet today. However other devices are rapidly growing, devices that may be more convenient for some users (TV sets), more portable (mobile phones) and much smaller (PDAs). We discuss about these devices in Chapter 6, however we consider here the impact that this trend may have in the software industry. As other devices are gaining access share, the value of controlling the OS for PCs decreases, and complementary products may move towards the new devices. Aware of this trend, the software industry is becoming involved in new battles to set and control standards in operating systems for mobiles, PDAs, game consoles, TV set boxes, etc. The industry may drastically change depending on the outcome of these changes. Will any new device dominate the others, or will all of them compete attracting different client segments? Will open or proprietary standards develop for each of these devices? The answer to these questions will describe a new scenario for the future of the software industry.

Another factor that affects the entire arena is a company called Microsoft. This company is a force in all three segments within this stage as well as a growing number of stages elsewhere along the value chain. As is well known (and considered by the US antitrust courts), Microsoft is the modern master at leveraging powerful assets and flattening competitors. It possesses huge financial, structural, sales and marketing advantages; it dominates the world in PC OSs and maintains a burgeoning position in server operating systems; and it strives either to partner with or simply to swallow up any threatening newcomers. Not only does Microsoft dominate the fundamental paths of IT technology, but (so long as US courts tolerate such incursions) aggressively pushes into promising markets in businesses it is not yet in, such as home mortgages (perhaps through a Money Advisor applet). It leverages its dominance in the major component of step of the value chain – OSs – to gain entry.

The future of the software industry will clearly depend on the result of the interaction between Microsoft and its powerful competitors, such as Sun Microsystems, Oracle, IBM and HP.

Notes

1. The following discussion is based on H.R. Varian and C. Shapiro, *Information Rules: A Strategic Guide to the Network Economy*, Boston, MA, Harvard Business School Press, 1999.
2. Varian and Shapiro (1999).
3. Initial Public Offering.

CHAPTER 3

Local Loop, ISPs and IAPs

This chapter analyses two steps of the value chain that have traditionally been considered in isolation; in terms of Figure 1.3 (p. 11), these are: (6) local loop operators (usually local telecommunication companies, 'Baby Bells' in the USA), and (5) Internet Access Providers (IAPs). As we will see in the chapter, these two steps are intimately related, not only because IAPs must use a local loop to reach their customers, but also because technological advances in the latter will completely restructure the IAP industry in the next few years.

3.1 The ISP/IAP Industry

An Internet Service Provider (ISP), also called an Internet Access Provider (IAP), is a business that provides services to connect individuals and companies to the Internet. An extension of the ISP is the online service providers (OSP), such as the online division of AOL Time Warner or Microsoft Network (MSN). The OSP provides an *integrated offering* by combining Internet access with a portal that includes exclusive and proprietary content. ISPs generate revenue mainly by charging user subscription fees, and may receive significant advertising revenue as well. To carry out our analysis of the industry we will study these companies as providing two different services, connectivity and information. In this section we will focus on 'pure' ISP companies that simply provide access to the Internet. The content part of the business for those OSP that provide it will be discussed in Chapter 8.

Currently, consumers have several ways to connect to the Internet – DSL,[1] cable and wireless are among the options – though most still connect through the 'Plain Old Telephone Service' or POTS (see Table 4.2, p. 67). The POTS method entails the following sequence: (1) the user dials the ISP over a modem and sends a data request over the telephone line, (2) the ISP then sends the data request to the appropriate server, (3) the server sends the requested data back to the ISP and (4) the ISP sends the data to the individual, requesting it via the traditional copper wire. When using traditional phone lines to dial up the ISP the user establishes a phone call and is subject to the costs of such a connection in addition to any the ISP might charge for accessing the Internet.

ISPs: Competitive Factors

Profitability

Only a small percentage of ISPs are profitable, for a number of reasons. First, *barriers to entry* are low because of the relatively inexpensive infrastructure required to start providing access. However, as the firm grows, income will be outpaced by the cost of expanding the necessary infrastructure. Acquiring users becomes expensive, requiring large investments in advertising to build brand. Secondly, *switching costs* are low because there is little opportunity for pure ISPs to differentiate themselves, aside from providing better service or higher-speed access. However, these strategies are not sustainable: the commodity status of access has led to a price war. Lastly, since cost structures are mostly fixed, companies have a tendency to *reduce prices* to increase client base and contribute to offset these fixed costs. Competition has led to massive consolidation. In Spain, for example there were more than 500 ISPs in 1995 and by 2002 they had been reduced to almost 10 per cent of this amount.[2]

Advertising Revenue

Some ISPs have adopted an advertising revenue model, providing free Internet access for consumers who are willing to provide personal information and permit a permanent space on their screens for advertising. Although free ISPs spend much less on acquiring customers, the amount they generate in advertising revenue is still lower than the actual cost of operating the network.

Economies of Scale

Economies of scale enable the firm to purchase access at lower rates and, if it is part of their business model, to attract more advertisers. Economies of scope enable the firm to capture the maximum profit possible from each subscriber. The success of bundling access with additional services depends on the quality of the customer base and the likelihood of that customer base wanting other services. Customers attracted to the free ISP model, for example, may not be willing to pay for additional services and may not be attractive for advertisers.

Size

In all of this, size is the critical factor. Without significant scope, few players will survive. In the USA, although only 20 per cent of ISPs operate on a national level, those who do generate over 80 per cent of total revenue.[3] Advertising revenues and returns from external commerce are highly concentrated among the larger providers, creating a cycle in which the large get larger and the small get smaller. This process can lead only to further consolidation within countries and even across entire continents, such as Europe.

Sources of Revenue

Seventy per cent of total ISP revenues are derived from consumer access subscription fees; so far, consumers have not shown a willingness to pay extra for value-added services such as email. The remaining 30 per cent of revenues is generated by expanding into hosting for businesses that are migrating to the Web. ISPs that focus on the consumer are moving toward offering multiple access technologies, ranging from wireless and DSL to cable.[4] Alternative sources of revenue will continue to increase in importance, leading ISPs to converge with other stages of the network. Simply speaking, ISPs are becoming OSPs – portals with Internet access.

Government regulation is important in determining the revenue model that ISPs pursue. In the USA, consumers pay a flat rate for local phone calls regardless of how many local calls they make (or how long they are). This is referred to as an *unmetered service*. In Europe, however, local calls are metered; therefore, on top of a flat monthly rate, users are charged based on the total minutes of local calls made. As a result, ISPs in the USA generate revenue by charging a monthly access fee, usually around $20, while in Europe, ISPs provide free access and generate revenue by taking a percentage (from 5 per cent to 25 per cent) of the local calls made to access the Internet.

Access Methods

The methods by which users access the Internet have been moving away from wired systems. In 2000, the ratio of mobile telephone users to desktop Internet users was 2:1. By 2005, when the number of global mobile phone users is anticipated to top 1 billion, the ratio is expected to rise to 3:1. The number of mobile telephones users exceeds the number of desktop Internet users in all regions except the USA. Even with the revised forecasts of 2002, half of all mobile telephones are expected by 2005 to be Internet-enabled,[5] providing a continuous seamless wireless Internet connectivity combining low-range LAN access and mobile phone-based access. Following a model pioneered by NTT DoCoMo, mobile network operators will want to play a greater role in the supply of value-added content as they will be able to offer their customers the ability to access all their information needs, irrespective of location, from one single service provider (Exhibit 3.1). But this will involve daunting capital expenditure. For one thing, to handle the traffic with 3G technology new transmission base stations will have to be built, about twice as many as there are now. Indeed, the cost of upgrading one of today's USA-based mobile networks to deal with the new technology could surpass $1 billion.

Exhibit 3.1 NTT DoCoMo

NTT Mobile Communications Network (commonly known as DoCoMo[1]) was born in 1991 when Japan's former monopoly telephone operator, Nippon Telegraph and Telephone (NTT), formed a subsidiary to manage their wireless segments that included: paging services, car phone services, in-flight phone service and mobile phone service.

In 1993 NTT DoCoMo launched their first digital mobile phone service. The market in Japan was liberalized a year later, enabling customers who had previously had to lease mobile phones from the network operators to buy them at retail stores. This, combined with the launch of competing personal handy phone services (PHS), triggered unexpected growth in the market, in which DoCoMo was the principal player with a market share of almost 50 per cent. By 1996 NTT DoCoMo's mobile subscribers reached 8 million.

In February 1999, NTT DoCoMo launched its I-MODE service, a high-speed mobile data service using specialized handsets (Figure 3.1). The project was extremely successful, attracting almost 1 of 4 Japanese by March 2002, and making it the most widely used mobile internet service in the world.

Figure 3.1   The I-MODE handset

I–MODE Model

For the customer, I–MODE consists of a simple cellular phone
furnished with a smallish screen that can display text and small graphics and that can typically hold up to 8 lines vertically. The phone can
be used like any regular cell phone for making voice calls, but by
pressing special 'I' button (see Figure 3.1), the user can be logged on
to a central gateway server operated by DoCoMo.

With I-MODE, cellular phone users get easy access to more than
62,000 Internet sites, as well as specialized services such as e-mail,
online shopping and banking, ticket reservations and restaurant reservation advice. Users can access sites from anywhere in Japan, and at

unusually low rates. The service was launched in a context in which the Japanese market for mobile phones was reaching maturity and users were in need for new services.

I-MODE uses a packet switching technology that enables NTT DoCoMo to charge by amount of data transmitted rather than per minute. It also permits an always-on connection to the mobile Internet, rather than dial-up access. The service makes use of its own proprietary cHTML language. The similarity to HTML permit programmers to build I-MODE compatible sites and services alongside their web-based products.

To complete the puzzle NTT DoCoMo had to gather appropriate content for the service. They sought a variety of high-quality mobile data services and applications like: financial information, travel, news and entertainment. The objective was to leverage their large subscriber base and dominant market share, building a 'virtuous circle' in which I-MODE quality content attracted more subscribers, and as more subscribers signed on, more content providers wanted to produce I-MODE sites and services. Although I-MODE was aimed at multiple age targets, the youth market – primarily users between 19 and 34 years old – has more than half of the subscribers. The most-used services are i-mode mail, mobile banking and train timetable and through-ticketing information.

The NTT DoCoMo pricing policy offered a low barrier to entry to consumers: I-MODE users paid a basic I-MODE subscription and being a pay-per-use fee, the average bill for i-mode data transmission about €12 per month. The company controlled all the billing for the I-MODE service, making its money from traffic revenues and from a 9 per cent commission payable by partners on all billable services (Figure 3.2).

I-MODE Success

One of the main factors in I-MODE's success was the size and dominance of NTT DoCoMo in the market place. The company offered a discount on calls between its own subscribers. Emails were exchanged substantially quicker when both parties were NTT DoCoMo subscribers.

Another factor cited to explain the success of I-MODE in Japan was the lack of penetration of wired internet usage, due to the low levels of home PC ownership in Japan, the high internet access cost and the lack

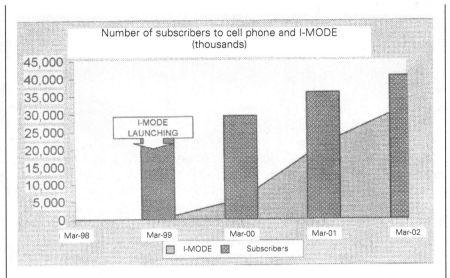

Figure 3.2   Subscribers to cell phone and I-MODE, March
1998–March 2002

Source: NTT DoCoMo 2003

of local language content. Cultural facts also contributed to the success
of I-MODE. Long commutes on crowded public transport provided an
opportunity for I-MODE usage. It was also considered unacceptably
rude in Japan to have loud conversation on mobile phones in public
places, possibly leading to an increase in the use of I-MODE-based
emails in lieu of voice calls. The Japanese predilection for fashion and
trends, and their fascination with gadgets are also cited as factors of I-
MODE success.

Finally, the marketing concept: 'I-MODE is a set of useful things
that can be done on a cellular handset' contributed to the adoption of
the new internet services of NTT DoCoMo.

DoCoMo's success was largely built upon the runaway success of
the i-mode service in Japan, the world's first mobile internet service. I-
MODE gave NTT DoCoMo the reputation of the world leader in
mobile internet technology.

Note:
1. DoCoMo is a Japanese word that means 'anywhere'.

ISPs: Strategic Approaches

Value-Added Services and Basic Access

An ISP needs only some basic connecting equipment located in a locale usually called a POP (Point-of-Presence). A POP in a city permits customers from that city to use that ISP to establish a local telephone call. An ISP like AOL spanning the whole of the USA needs a POP in each local phone area in order to avoid its clients having to dial-up long distance. A small ISP will then lease a high-speed line from a traditional telecom operator to link its POP with an Internet Access Node. Larger ISPs often have their own high-speed networks; they are then less dependent on the telecom suppliers and can offer better service to their customers.

In this straightforward stage of the network, a company connects the caller to the Internet, not bothering to identify that caller commercially. An ISP is then simply a *bridge*, using its hardware to get to a phone line. Yet its function is essential; without ISPs, the Internet wouldn't exist. But by itself, an ISP is not a substantial business proposition. Therefore, competitive strategy primarily involves mixing in value-added services to basic Internet access. The goals are to differentiate the individual provider, to create customer-switching costs and to incorporate profitable services which both grow and profit from the installed base.

Providing Exclusive Content

Such is what AOL believed it was setting the stage for when it acquired Time Warner. To sustain its competitive advantage, AOL felt it had to be in a position to provide exclusive content. Putting the theory into profitable practice proved much tougher than AOL supposed. The company reported a hefty loss in April 2002, with its stock plummeting to less than $18 a share – down some 66 per cent from merger-day in January 2001. It then replaced its US search engine, Overture, with Google (formerly GoTo), hoping to get a revenue boost from the advertising the latter's paid-listing advertising (in which companies pay to be included in search results) was expected to bring in. Despite a slowdown in its customer growth rate, America Online has remained far and away the largest ISP worldwide, and TimeWarner the largest content provider.

Access Prices

Access prices have fallen in countries outside the USA, and unmetered telephone billing seems destined to become the standard. In 2000, AOL forced British Telecom to offer unmetered access to AOL (and as a regulatory consequence, to its competitors as well) throughout the entire country.[6] That, in turn, meant that ISPs would be able to offer fixed monthly rates to consumers without having to bear extra charges tacked on by phone companies.

ISP Functions and Broadband

Taking advantage of their brands and their relationships with customers, the telecoms themselves have established their own ISP functions. Although additional revenues from access alone may prove to be minimal at best, the step opens the door to new business areas that can spawn fresh income sources, such as access to the Internet through mobile phones. This trend has produced a pool of providers, which greatly increases the level of competition.

Along with telephone firms and ISPs, cable and satellite companies have become contenders in the race to deliver broadband service – although in 2003 the consumer demand for broadband was not as strong as expected. (These technologies provide a broadband 'last mile' that can be used to attack the monopolists' conventional narrowband local loop.)

ISP/IAP: Industry Structure and Trends

Wireless-LAN

ISPs face enormous difficulties in capturing any value from their customers. As we have already described, low entry barriers, low switching costs and an extremely commoditized product are structural characteristics that impede sustained profitability. Technological advances also make companies in this industry vulnerable to sudden obsolescence. An example of these impending changes is the vast proliferation of Wireless-LAN access at the customer level (see Exhibit 3.2). This in itself should not change the industry, since it is already common for an end-customer with broadband access and a wireless router in his or her home to allow the neighbours to hook up without any impediment. If this practice becomes commonplace, the ISP industry will see its final subscriber pool greatly reduced, as many individuals will share the same access point.

'Hotpoints'

A second interesting trend is the vast proliferation of public access 'hotpoints'.[7] Many public establishments provide wireless internet access to their customers as an additional service. Again, if one can find complimentary access in bars, bus and train stops, airports and bars, the IAP industry will without a doubt be reshaped (see Exhibit 3.3).

Exhibit 3.2 Wireless-LAN: Wi-Fi, the 802.11B Technology

Wi-Fi is the short for *Wireless Fi*delity, and is meant to be used generically when referring to any type of 802.11.network, a family of specifications developed by the IEEE (Institute of Electrical and Electronic Engineers). The term is promulgated by the Wi-Fi Alliance, an organization made up of leading wireless equipment and software provider with the missions of certifying 802.11-based products for interoperability. Wi-Fi technology allows you to connect to the Internet from your home, a hotel room or a conference room, because it is a wireless technology like a cell phone, that enables computers send and receive data indoors and out, anywhere within the range of a base station, and it is several times faster than the fastest cable modem connection. In a typical setup, laptop users tap into the network via a so-called 'access point,' which is a radio transceiver mounted on an office ceiling and connected by cabling to the wired network. One access point usually supports between 12 and 20 people. (It can handle more, but congestion becomes an issue.) All the user needs is a network interface card (NIC) either embedded in a laptop or inserted into a slot in the machine.

The Range in a Wi-Fi network varies depending on the type of Wi-Fi radio you have, whether or not you use special antennas and whether your network is in an open environment or in a building. The data transmission rate depends on the distance between the access point and the PC or the device to connect to the Internet, and normally ranges from 1mbps to 11mbps. When you move farther away from the access point, a gradual degradation in range occurs instead of stopping all together, your data transmission rate becomes slower.

Wi-Fi is becoming one of the most useful technologies to access the

Internet, which is why companies, governments and other organizations are beginning to adopt it. McDonald's, for example, and Soft-Bank plan to create almost 4000 hotspots in the McDonald's restaurants in Japan to give its customers wireless broadband access to the Net. The German air company Lufthansa has carried out a first flight with the Internet connected on board using Wi-Fi technologies. By 2004 almost all airline companies will offer Internet services in their flights using the same technology. In Spain, Zamora has become the world's first city to install a broadband wireless internet access service that provides coverage within the entire city limits. Some 150–200 wireless access nodes will be installed to cover all Zamora; users will need to subscribe to the ISP and install a Wi-Fi card in their PCs. In the labour market, workforce trends are also driving Wi-Fi deployment. Knowledge workers are spending less time at their desks and more time in conference rooms and training seminars, or are working from home. In the IT industry, Intel, IBM and AT&T have founded a new company called Cometa Networks, which plans to install more than 20,000 wireless broadband Internet access points, or 'hotspots', around the USA by 2005. Cometa's goal is to provide residents of the 50 largest metropolitan areas of the USA with wireless Internet access points that are no more than a minute walk away. The most amazing example of the proliferation of Wi-Fi technologies is a Cyber Café located at 5,000 m in a base camp on Everest!

Wi-Fi Technologies and so have some disadvantages. Wireless-LAN has much less privacy compared to a wired LAN because of the use of a radio frequency (RF) as a transmission medium. Anyone with compatible device can receive the RF transmission in its range and connect to the LAN server. Another disadvantage is that the data transmission rate can fluctuate significantly in comparison with a wired LAN.

Exhibit 3.3 Starbucks' 802.11B 'Public' Access

Starbucks coffee began as a grocer of roast coffee in 1971, opening its first location in Seattle's Pike Place Market. In 1982 Howard Schultz joined Starbucks as director of retail operations and marketing, by this time the company provided coffee to top restaurants and espresso bars.

In 1983 when Schultz travelled to Italy, he was impressed with the popularity of espresso bars in Milan and he saw the potential to develop a similar coffee bar culture in Seattle. In Milan, the coffee shop was the principal place to meet people. When Schultz went back to Seattle, he tried to convince the founders to test the coffee bar concept in a new location in downtown Seattle. Although he persisted with the idea he did not succeed. In 1985, with the backing of local investors, Schultz acquired Starbucks assets and changed its name to Starbucks Corporation.

The main objective of Schultz was to redefine the concept of Starbucks' business: 'Coffee is not a product, nor a service, coffee should be an experience,' he said. The Starbucks experience is about passion for a quality product, excellent customer service and skilled people. This strategy worked well and the company now has more than 6,000 stores in 25 countries, with some 20 million customers.

In 2001 the company launched a nationwide campaign to put wireless access with T1[1] speed in some of its stores, with Wi-Fi technology. Starbucks signed a deal with T-Mobile, the wireless subsidiary of German-based Deutsche Telekom, and with Hewlett Packard. T-Mobile offered Starbucks' customers a free trial of its T-Mobile HotSpot service for 24-hours, after which the company would charge ISP prices. In addition, T-Mobile offered a variety of Internet access service plans, including National and Local Unlimited monthly subscription plans, as well as Prepay and Pay-As-You-Go plans on their Web site, at their stores and at Starbucks. HP's real contribution as in its new Wireless Connection Manager. The free, downloadable software made it simple for mobile users to configure their notebook or PDA. The software was available as a free download.

The principal reasons to launch an internet wireless service at a coffee shop was that 90 per cent of Starbucks customers usually surf the Net and a high percentage of them used to go to Starbucks to finish their work or home task, while they drank a Cappuccino. Now with this new service, a customer could check their email, set a chat and surf the Net while is drinking a coffee. To connect to the service, customers needed a T-Mobile HotSpot account and Wi-Fi capability for their notebook computer or Pocket PC. The next step was to download special HP software called Wireless Connection Manager, enabling users to configure their notebooks and Pocket PCs to automatically sense and connect to available wireless networks. With the software and hard ware already installed, the customer had only to turn on the computer and sign in T-Mobile Network to start to surf the Net.

The Wi-Fi service is the beginning of a more ambitious project that includes the development of e-commerce services, in which the customer makes an order and pays electronically through her hand-held or cellular phone; the provision of exclusive content about coffee, music and other entertainment; The use of the network to improve the efficiency of Starbucks' shops and the back-office of the company. The service costs users 20¢ per minute or $15.95 per month, while using software and services from Microsoft. This is more expensive than other Internet access alternatives, but Starbucks' executives believe that the benefit of a speedier connection justifies a more expensive fee.

Note:
1. A dedicated phone connection supporting data rates of 1.544Mbits/second. A T-1 line actually consists of 24 individual *channels*, each of which supports 64Kbits/second. Each 64Kbit/second channel can be configured to carry voice or data traffic. Most telephone companies permit you to buy just some of these individual channels, known as *fractional T-1* access.

Wireless Routers

A third trend and probably the most disruptive for established IAPs is the deployment of public access wireless networks using the same wireless LAN 802.11 technology. Companies are installing thousands of wireless routers in telegraph poles, buildings and even trees in parks. The idea is to provide wireless access to a whole area without the need to deploy any cables: with the appropriate authentification and billing systems, a company could provide Internet access to a whole city without requiring any additional infrastructure. Wireless routers are extremely cheap when compared with the antennae and infrastructure necessary for 3G cellular telephony, and compare very favourably with the costs associated with DSL land-based access.

3.2 Local Loop

DSL and ADSL

In telephony, a 'local loop' is defined as the connection from a telephone company's central office in a locality to its customers' telephones at homes and businesses. The local loop has traditionally been – and in most countries

still is (even in the light of deregulation policies) – a monopoly business for
the telecom serving the given locality. This also is referred to figuratively
as the 'last mile,' because it represents the final link from the backbone
infrastructure to the end-user. The connection is usually made through two
copper wires called a 'twisted pair.' Telephone companies commonly run
twisted pairs of copper wires to each household. The pairs consist of two
insulated copper wires twisted into a spiral pattern. Although originally
designed for the 'plain old telephone service' (POTS, see Table 4.2, p. 67),
these wires can carry data as well as voice. The traditional modem transforms the computer's digital data into voice-like signals so that the existing
voice system can transport the data from computer to computer. At the
receiving end, another modem transforms the voice-like signal back to the
original digital data. New services such as ISDN (Integrated Services Digital Network) and DSL (Digital Subscriber Line) also use twisted-pair
copper connections but they do not resort to tricking the system by transforming digital data into analogue voice-like signals. Of these new technologies, DSL and ADSL (Asymmetric DSL) in particular, merit special
attention. ADSL allows ISPs to provide high-speed Internet access to
customers over the traditional twisted copper pair without interfering with
the normal voice-carrying operation of the line. Even in countries where the
local loop is protected by monopolistic regulation, DSL has permitted new
companies to offer high-speed Internet access without having to invest in
the deployment of an additional local loop. Other local loops include cable,
the electrical wiring and direct satellite. With this last technology, the user
downloads from the satellite at amazing speeds (up to 25
megabytes/second) and uploads through the phone company at traditional
modem speeds. This is called a 'mixed local loop'.

Commercial Value

The local loop is the most commercially sought piece of the communication infrastructure, as the direct relationship enables the telecom who owns
it to capture customer information, including demographics, billing, and
knowledge of time spent online, providing telecoms with an edge in offering value-added services.

Local Loop: Competitive Factors

Unbundling

The ability to make money in the local loop depends only on the local competition on the loop. A fundamental factor in this value-chain stage is the liberalization of the telecom sector in the late 1990s that opened the local loop to competition (a deregulation process referred to as 'unbundling'). Different countries are at different stages in the unbundling process, with the USA leading the way in 1996; Europe started in 2001, with pressure increasing in many countries to admit competition.

'Sweet Deals' and Churn Rates

Despite widespread deregulation of telecoms worldwide, former monopolies are still resisting giving up their control over the local phone network and their protected relationship with the end-user. This situation leads to often self-defeating price wars. How else do you differentiate the first phone company from the second phone company? In Spain, for example, there is little money to be made by entering local-loop competition, because there is a countrywide monopolist, Telefónica. In Valencia, however, a cable company decided to deploy an alternative local loop, and a large fraction of telephony and Internet-access customers switched to them.[8] But to do so, the cable company had virtually to give away its local-loop service, because they could not offer the customer at the end of the value chain anything that was significantly different from Telefónica. With undifferentiated service and customer switching costs above zero, new entrants have to offer 'sweet deals' that will tempt clients from the incumbent to the new company. This sets the stage for retaliation by lowering prices and incentivizing the switch back, leading to high churn rates and lower ARPUs (Average Revenue Per User). Multiple-access loops with comparable technologies permit us to model the land-line local operators to cellular companies, which have so far been plagued by very high churn rates given their inability to avoid galloping customization.

Excite@Home

As has been shown in the USA by, for example, the failure of Excite@Home, there's little chance of making money as one of two alternative loops vying to serve a single locale. The problem is that when two companies go against each other it's not that each takes half the pie, but that the pie actually shrinks! Excite@Home faced the problem of trying to differentiate its capabilities from the competing telecom. To do so it offered vast bandwidth. But as a cable company providing TV, Excite@Home proffered scant content to fill all the Internet bandwidth. Such infrastructure is expensive (it leased AT&T cables to provide Internet access) and, at lower cost, an existing phone company can sell a customer DSL access on the existing local loop; although it is perhaps half what the customer might realize through broadband, that is acceptable to most users given the lack of high bandwidth-requiring content on the Net. And at any given time, the phone company can mount a price-dumping campaign (where it's within the law), and the competing company will find itself in deep trouble.

Costs and Economies of Scale

Those who would nonetheless compete in the local-loop stage will find that technological advances play a central role. Because business and residential customers expect greater bandwidth, speed, service and flexibility, players must offer broadband and other advances. This creates high barriers to entry owing to the large investments needed to provide the transmission lines. Bringing fibre cable to homes or office buildings entails the relatively large labour costs of digging up streets to lay the cable. Above the road, wireless transmission requires erecting regional towers. Because such large investments are needed, economies of scale are important; this leads to fierce competition to grow installed bases. Incumbent telecoms benefit from already having a physical wire connection to the home (though in many cases it's only old-fashioned twisted copper).

Some cable companies also have a data-carrying infrastructure in place, especially in the USA, where a plethora of cable companies have connected cable and affiliated broadband to millions of homes. Technologically speaking, that connection usually is much better that the phone company's. It can be fibre optic to the kerb, and handle large numbers of megabytes. But do customers really need it? And what price can they be charged? A phone company can install a DSL next to its voice wire and offer 1 Mb at a decent price. How much can a cable company expect to get for 2–3 Mbs if few customers need that much?

Local Loop: Strategic Approaches

Incumbent operators respond to attacks on their once-protected markets by upgrading and expanding existing infrastructures, which otherwise couldn't keep up with rising demand. Their goal is to maintain their customer base by making optimal use of their local loops – incorporating broadband technologies that combine with, or in some cases, replace the copper wires of the traditional infrastructure – in order to be able to offer an expanded line-up of services. These modernized capabilities are then either utilized by the incumbent to roll out profitable value-added services built upon flexible and scalable platforms, or resold to other service providers. In Spain, for instance, Telefónica is selling complete DSL connectivity to companies that resell the service under their own brand to final customers (Exhibit 3.4). In general, when there is a dominant incumbent player, the wholesale price has an upper bound set by regulators. In many countries, including the USA, local phone companies rent space in their premises for new entrants to install their DSL equipment. In this instance, the new company has to invest in the equipment and install or lease lines form the incumbent premises to an Internet access node.

Exhibit 3.4 Telefónica's Wholesale IAP Business (vs. Terra as Client)

Following the liberalization of the Spanish telecommunication market that begins in the late 1990s, Telefónica, the former Spanish monopoly of telecommunication, was compelled to open the local loop to new companies who could offer internet and broadband services to the end customers of all the Spanish market. Telefónica decided to play in both, the wholesale and retail internet access markets.

As a wholesaler, Telefónica has developed three business models to cater the new entrants: the first one, a Basic Level, is a contract in which Telefónica offers to lease out to the Telco operator its copper lines, which connect each of Telefónica's local exchanges with the end-user. In this contract a high investment of the Telco is required to equip each of the hundreds of exchanges located around the country with its own technology. This business model did not work because each Telco operator had to invest a huge amount of money in a market in which paybacks were not guaranteed. So Telefónica developed two new business models with significant differences in the investment required. In the first one, called *GigADSL*, Telefónica divided the country into geographical territories, offering the Telcos to group together all of a customer's end-user traffic in each territory and then funnel it directly to the customer's central operation under its administration. Choosing this model the Telco can save money in investments, because its has only to develop the infrastructure in the district.

The second one, called *Virtual ISP* is a model in which Telefónica offers its telecoms networks for all the operations of the Telco, so the end-user is no longer technologically dependent on her ISP for the

connection, and the Telco practically has no requirement for investment, but the fee charged by Telefónica for this service is higher than
the one charged for GigADSL service.

In the retail market, Telefónica functions as an ISP, providing
analogue narrowband and digital broadband services, offering
customers Internet access and content. In Spain, in the broadband retail
market, the ADSL is the most prevalent technology. Telefónica begin
to play in the ADSL market, through its controlled company, Terra
Networks, that rapidly become the most popular Spanish Internet
portal and the largest Internet service provider.

But in September 2001, Telefónica decided to change the strategy and
began to offer ADSL to the end-users by itself, competing against Terra
and other ISPs. There are many reasons to explain this change. First the
analysts predicted that one of the most profitable telecoms business
would be to provide broadband Internet access. Another argument was
that Terra was losing market share in the ADSL market, so Telefónica
wanted to reinforce its presence. If Telefónica and Terra competed in
the ADSL market, Telefónica's Group would win, because the companies has different targets: Telefónica focused principally on corporate
services and small business, while Terra, in the business of internet
contents, was well known by Internet surfers. During the first 4 months
of selling ADSL, Telefónica doubled the customers base of both Terra
and the other ISPs.

In February 2003, the strategy of the Telefónica Group changed
again. Terra was to be the Internet content and email provider for the
holding company, and the reseller of Telefónica's ADSL service, and
would outsource the management of certain services and network
elements to Telefónica Group companies. Telefónica would pay Terra
a fee during the next 6 years for the content service. This strategic
alliance took full advantage of Telefónica's capacity as a broadband
and narrowband connection and access provider, and Terra Lycos'
position as a portal. The aim of this agreement was to take advantage
of synergies and create value for both companies.

Alternatives to DSL and cable modem connections are fixed wireless,
either traditional radio or optical. A small antenna is sited on the roof of
each building or house within range of a transmission tower. The wireless
local loop thus created enables them to bypass incumbent local-loop
networks, and therefore avoid access fees as well as the high costs of laying

cable. In addition to being very scalable, bandwidth available on radio frequencies is exceptionally high, second only to the nearly limitless bandwidth of fibre optics.

Local Loop: Industry Structure and Trends

This value-chain stage is closely related to its surrounding stages, backbones and Internet access. Companies – especially telecoms – competing within these stages are not only integrating and leveraging services among and across the three markets, but, perhaps to a lesser degree, as we will see later, are also combining services from other stages as well, such as online content, hosting and portals. Country-by-country regulation and deregulation will continue to play a defining role in this stage.

After the appearance of DSL that has allowed competitors to enter the IAP industry using the existing local loop without interfering with voice operations and therefore making unnecessary to invest in additional local-loop infrastructures, the next big revolution will be the wireless local access. In contrast to 3G cellular, wireless local-loop technologies use open frequencies and do not require the approval of the local governments. Companies may enter the industry at different levels. The most common practice is to provide Internet access to individuals and companies without paying fees to local telecoms, but some companies are starting to deploy infrastructure without entering the IAP business, but instead reselling access to IAPs. This business model is hardly sustainable. We have already discussed the difficulties of IAPs with traditional technologies; when a whole set of access alternatives appear, competition will increase even further, lowering profitability opportunities even more.

Notes

1. DSL (or Digital Subscriber Line) is a technology that uses the same wiring as traditional telephony to transport information in its original digital format. With DSL, the wiring can be simultaneously used for conventional voice communications. See the Local Loop section on p. 00 for a more comprehensive definition.
2. The Spanish Association of ISPs (AEPSI) stated that at the end of 2002 there were less than 50 ISPs with a profitable business. By this time, the offer was concentrated in only 15 players.
3. CyberAtlas.com, 'National ISPs Still Kings of the ISP Hill', September 28 2000.
4. Cahners In-Stat Group, 'ISPs to Generate US $32.5 Billion in US', September 27 2000.
5. CyberAtlas.com, 'Wireless Aims for Widespread Appeal', February 13 2001.

6. D. Einstein, 'Flat-Rate Net Access Finally Arrives on British Shores', *Forbes.com*, 25 September 2000.
7. See http://www.80211-planet.com.
8. Cable Europa SA, under the brand of ONO, headquartered in Valencia, is now the largest cable operator in Spain. It offers direct access telecoms, cable television and high-speed Internet access to residential and business customers in four large geographical clusters around Spain. In Valencia at the end of 2002, 27 per cent of telecom residential services and 8 per cent of Internet access were provided by ONO.

CHAPTER 4

Network Infrastructure: The Internet Backbone

4.1 Infrastructures: The Bandwidth Revolution

Bandwidth is an expression of the speed at which digitized data can travel over a conductor such as a telephone wire (relatively slow) or a fibre optic cable (relatively fast). We shall define 'bandwidth' simply as the maximum amount of data in megabits per second (mbps) that can be sent from computer A to computer B, thereby expressing the capacity of a networked connection. In essence, the more bandwidth there is, the more data that can travel along that connection within a certain period of time. As Table 4.1 shows, some applications require very little bandwidth, others quite a lot.

Fibre-optic cable is already able to carry transmissions at 9.8GB per second (9,800,000,000 bits a second). At that speed, one computer could

Table 4.1 Size and bandwidth

Bandwidth necessary to download a file from internet in 1 second

Type of file	Observations	Size (kilobytes)	Bandwidth required megabites) per second)
Text file	File of 20 lines	2	0,0
MsWord file	Typical Docs with chart and images	110	0,9
High-quality picture (.Tiff file)	An image scanned from a 10 x 15 cms photograph	6.000	48,0
High-quality picture (.JPEG)	An image scanned from a 10 x 15 cms photograph	650	5,2
DataBase file	An MsAccess database file	2.000	16,0
MP3 file	Song of a duration of 3 minutes	3.500	28,0
AVI file	High-quality video of a duration of 3 minutes	9.000	72,0
Content of a floppy disk	N/A	1.400	11,2
Content of a DVD	N/A	3.675.000	29.400,0

transmit an entire DVD to another computer in 3 seconds. (The theoretical limit of fibre-optic cabling stands at the once-unimaginable pace of 100TB per second (100,000,000,000,000 bits per second)!

The bandwidth revolution has not taken place in isolation. Its technology has been intertwined with increments in computing power and the compression of data. That combination, which packages data in ever-swifter-travelling bundles, has enabled bandwidth to develop more rapidly than it would have on its own.

Yet some predict that the expected surge in bandwidth power will render the networked computer obsolete as an information-gatherer. Whether that conclusion is true is not yet clear. But that has not stopped the world's futurists from speculating like science-fiction writers. 'Simply put', technology guru George Gilder has already claimed, 'bandwidth is a superior substitute for computer power, memory, and switching. It can handle far more data, far faster, and with far fewer problems'.[1] Gilder has coined the term 'telecosm' to describe the universe of instant-data accessibility which he predicts information technologies will inhabit:

> The mind-boggling amount of data and the speed at which it will travel via bandwidth technology will render today's time-wasting practices both obsolete and detestable. The Internet will be the conduit. Fiber-optic networks, digital wireless, and cable modem will provide the access. Miniaturization and portability will make it practical. And the increasing value of time will make it imperative. Communication, banking, investments, shopping, news, weather, entertainment, education, travel booking, medical advice, buying a new car, selling your home, all this and much more will be at your fingertips via the telecomputer. This device will be as portable as a watch, as personal as a wallet. It will recognize speech, navigate streets, collect mail, conduct transactions . . . It will link to a variety of displays and collapsible keyboards through infrared or radio frequency connectors. Soon, bandwidth technology will usher in a wonderful new era of time-saving products and services. We need to look at a new era of physical connection devices – line switching, packet switching [see Exhibit 4.1], narrowband, wideband and broadband [Table 4.2].

Exhibit 4.1 Line Switching vs. Packet Switching

Line Switching

In a line switched data network, a physical connection is established by the switchboard through the communication network between two nodes (for instance two telephones A and B). This connection is used exclusively between these two subscribers until the connection is dropped. So in the use of this technology a fixed share of network resources is reserved for the call and no others can use those resources until the original connection is closed. One advantage of line switching is that it enables performance guarantees such as guaranteed maximum delay, which is essential for real-time applications like voice conversations. On the other hand, line switching is not good for transmitting data between computers: Every time a PC sends a file to another, a new connection must be established, so the process is inefficient. Moreover if all the lines are engaged, line switched networks refuse the new connection (Figure 4.1).

Packet Switching

The Internet uses 'packet switching'' technology. In a packet switching network all the PCs connected to the net are 'talking' each other all the time, and the data that will be sent are assembled into chunks of data called 'packets'. Each packet contains some information about the source (where it is coming from) and the destination (where it is going). A packet switched data network is made up of several computers called routers linked together. A router 'examines' each packet and

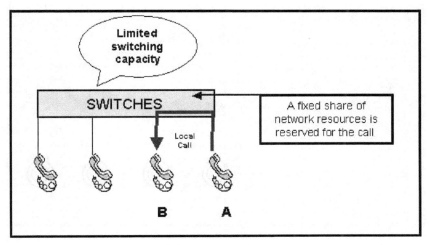

Figure 4.1   Line switching network

decides where to send it to move it closer to its final destination. Packets generated from the same data can travel different routes and arrive in a different order. At the destination, the packet switching exchange passes the packets to the receiving computer which makes sure that the packets are put back into the original sequence (Figure 4.2).

The main advantages of packet switching in comparison with line switching is that as the routers in the network are sending and receiving signals all the time, it is not necessary to make a new call every time you want to send information from one computer to another. The other advantage vs. the old technology is that the system is never busy. Packet switched networks can accommodate multiprocess communications, so never refuses a connection; at most, its can delay the connection until the packet can be transmitted. The principal disadvantage of packet switching is that packets can arrive at different times and in a different order than when they were sent: this is a problem for telephone conversation.

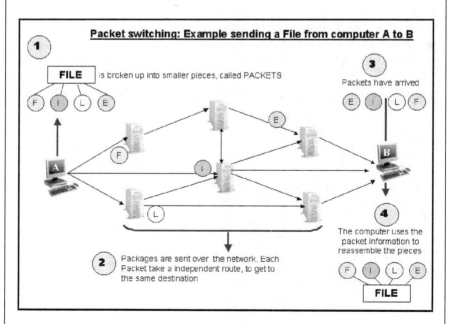

Figure 4.2   Packet switching: sending a file from computer A to B

Table 4.2 Narrowband, Wideband and Broadband connections

Narrowband connections: bandwidths of 64kbps or below

POTS	The 'Plain Old Telephone Service' Basic analogue voice telephony over a pair of copper wires	Can carry data at speeds of up to 56.6 Kbps with traditional modems
GSM	The Global Systems for Mobile communications is one of the leading digital cellular systems It is available in more than 100 countries and has become the *de facto* standard in Europe and Asia	Can carry data at speeds of 9.6 Kbps

Wideband connections: bandwidths between 64kbps and 2mbps

ISDN	The Integrated Services Digital Network is an international communications standard used over digital telephone lines or normal telephone wires	Offers one line with speeds of 64 Kbps, or two lines with speeds of 128 Kbps

Broadband connections: bandwidths of over 2mbps

ADSL	An Asymmetric Digital Subscriber Line Permits more data to be sent over existing copper telephone lines (POTS)	Supports data rates from 1.5 to 9 Mbps when receiving data (downstream), and from 16 to 640 Kbps when sending (upstream)
GPRS	The General Packet Radio Service. Permits GSM mobile phone operations to offer much broader bandwidth connections Often Referred to as '2.5G', since it is a step between GSM and 3G (see UMTS below)	Can carry data at speeds of up to 150 Kbps
UMTS	The Universal Mobile Telephone System is a third generation (3G) mobile technology This technology consists of a bundle of glass (or plastic) threads (fibres) that transmit data along the cable as light waves	Can carry data at speeds of up to 2 Mbps, Can carry data at speeds of many Gbps

This stage consists of companies that comprise the backbone – the network that connects one metropolitan area to another around the world for the Internet. It consists of three primary segments: (1) telecommunications companies (telecoms), (2) network equipment companies and (3) Web server hardware and software companies.

4.2 Long-Distance Telecommunications Companies

Traditionally, integrated companies had provided telecommunications that under regulation provided both local and long-distance services. Familiar names like AT&T in the USA or the nationally owned monopolies like France, Deutsche and British Telecoms, and the Spanish Telefónica ruled the industry. When deregulation allowed competition at different levels, a number of new companies entered the long-distance arena. Long-distance carriers require comparatively less investment than local carriers, as there is no need to string wires to final customers. These enterprises build and manage the infrastructure, as well as providing long-distance communication services. The backbone providers today are the former national telecom monopolies that owned the backbone infrastructure and a myriad of newcomers. Familiar big names in this sphere are Cable & Wireless, AT&T, GlobalCenter, Qwest, Winstar, Sprint, UUNET/Worldcom and GTE. These and dozens more phone operators act primarily as wholesalers, but in some instances also sell connectivity directly to end-users, particularly large corporations.

Traditionally the telecoms had been very profitable, controlling communications networks that have been a scarce and expensive resource in the regulated environment, but when deregulation happened and customers (both local operators and large companies) could choose from different alternatives, they faced a situation of competition that greatly reduced their profits.

Long-Distance Telecommunications: Competitive Factors

Overcapacity

Today the telecom industry is undergoing a massive transformation in which traditional telecom services (mainly, voice and data transmission) are becoming a commodity. As governments worldwide deregulate and privatize telecom services, many new companies are entering the market and building their own long-distance networks – from cable to fibre to wireless. Forecasts of huge data transmission capacity needs have not come to fruition, and the crude reality is that there exists overcapacity of transoceanic and other long-distance networks. In some areas of the world as little as 2 per cent of the installed capacity is in use, and the norm is that more than 50 per cent of the laid fibre is still 'dark' (has never been used). Advances in the software and hardware industries that permit faster compression algorithms and error checking effectively also increase the capacity of the installed networks. One could almost say that Moore's Law applies – that without doing anything to the existing network its capacity increases.

Higher-Margin Offerings

Due largely to this overcapacity, the telecommunications industry was among the hardest-hit stock group in the USA and European markets in the late 1990s and early 2000s, with the value of some companies effectively vanishing.[2] Although data traffic has been exploding at a rate that doubles every three to four months, carriers are struggling to turn this growth into profits. Incumbents have to overcome declining prices in their voice businesses, while newcomers have to find revenues fast enough to pay for their networks without setting off a destructive price war. Many data services are relatively inexpensive for consumers because telecoms have difficulties adding value to a transport service that is becoming commoditized. This is forcing the telecoms to pursue new higher-margin offerings, such as running virtual private networks, etc. Voice remains the profit-generating part of the business. John Roth, CEO of Nortel Networks, stated in 2000 that: 'The volume is in the Internet, but the money is in voice. Companies don't like to talk about that because the sizzle has been on the data side.'[3]

Acquisition and Switching Costs

Barriers to entry have dropped, due to deregulation and the fact that the growing digital business requires less expensive infrastructure than the traditional analogue voice business. The newcomers' data networks have strong technological advantages over their older rivals. Large installed bases are important for telecoms to profit from their infrastructure; once the capacity is in place, they compete aggressively to fill that capacity, as the marginal cost of serving an additional customer is extremely low. This causes customer acquisition costs to increase and switching costs to fall as customers are less influenced by who owns the networks and more concerned with who can provide the best solutions to their problems for the best price.

Revenues are also being pressured by the substitution of voice traffic by cheaper data traffic over IPs (Internet protocols),[4] such as email. In addition, government regulation for the development of the Internet together with the growing competition in the sector is forcing firms to accelerate the reduction in Internet usage prices. It is expected that the increase in traffic will eventually compensate for lower prices.

Telecommunications: Strategic Approaches

Upgrading Networks

Anticipating what was supposed to be an explosive demand, telecoms have spent billions of dollars to build or upgrade their networks to accommodate the expected growth in traffic. Another strategic response was to concede the 'transport only' business to the newcomers, who usually have cost and capacity advantages. This allowed the incumbents to focus on strengthening their retail interface with business and residential customers and on developing a more complete package of solutions that went well beyond just raw transport.

Higher-Margin Services

The complete solution involves adding the higher-margin services, such as virtual private networks, data storage and security, that can make the data business a profitable one. This in fact involves moving into other stages of the value chain, such as ASPs, ISPs, hosting and online content (see Figure 1.3, p. 00). An example of a successful telecom company that has done just this is the Spanish former monopoly Telefónica, that besides traditional voice and data communications, is offering virtual private networks, web hosting, support for online gaming and even computer games for rent. With these services, firms can capture a bigger share of the customer's total telecom spending. Only in this way can they create the economies of scope that provide an attractive return for the high customer acquisition costs.

Excite@Home

An example of a not so successful company was Excite@Home which tried to deliver a broadband connection between E-jo and the Internet but failed to create a compelling value proposition. Excite@home tried to solve the problem inherent to the packet switching distributed approach of the Internet (see Exhibit 4.1). The problem is that there is no way to ensure an end-to-end bandwidth given because any router may be overloaded. To minimize this, Excite@Home attempted to create a fast bandwidth Internet with would guaranteed speeds. To do so the company developed multiple links to the Internet (so as to minimize bottlenecks), replicating the content (so as to minimize traffic), using a proprietary fast backbone end to end and had its own hosting and local-loop arrangements (see Figure 4.3). However, the value proposition did not catch on, mainly for two reasons: first because of the deployment costs and second because there was no clear demand for broadband services.

Telecommunications: Industry Structure and Trends

The boundaries between different providers and services they offer are becoming blurred. Although many new telecoms have appeared with the

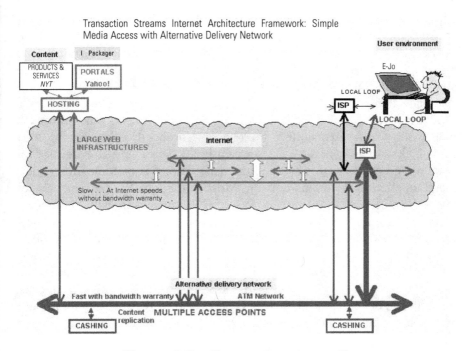

Figure 4.3 Transaction stream III

development of the Internet, it is expected that the sector will consolidate. The need for scale economies as well as pressure from the market for single-source, end-to-end providers is driving an ever-expanding list of telecom mergers, acquisitions and strategic partnerships among incumbent and newcomer operators as well as ISPs.

4.3 Network Equipment: Network Infrastructure

The companies in this category are dedicated to the design, manufacture and distribution of the network equipment – hardware and software – that telecoms and other businesses need to make the Internet work seamlessly. This equipment enables businesses to make connections to the Internet, as well as build Intranets (networks connecting people within an organization).

The Network Equipment Industry

The network equipment industry has been dominated by a handful of companies that has developed proprietary technologies. Even though the Internet and in some sense voice networks use open standards to communicate among themselves, the technologies of the actual switches and routers are proprietary to each manufacturer. Companies like CISCO in IP switching and Lucent, Nokia and Ericsson among others in data and voice dominate their markets. In early 2000 telecom companies were hit by a drop in demand and drastically halted their investments; the estimated worldwide capital expenditure for telecoms dropped 6 per cent in 2001 from 2000 and an incredible 32 per cent in 2002 from 2001. Forecasts for 2003 are not more promising, and the network and communication equipment suppliers were the hardest hit. Most of the telecom equipment sales made to non-telecoms were made to startup companies, mostly 'dot.coms', which were the hardest hit by the slowdown of the new economy frenzy. Equipment companies, and their investors, were used to double-digit quarterly growths that drove stock prices extremely high. CISCO, for instance, saw its shares rise (split-adjusted[5]) from $9 in January 1999 to $80 in March 2000. When growth stopped and sales actually shrank, inventories piled up and obsolescence crippled its book value, investors lost confidence and shares plummetted. In January 2003 CISCO stock traded at around $13.

Network Equipment: Competitive Factors

In contrast to the PC industry, entry barriers for network equipment are high. Several factors account for this. One is the technological innovation offered through fast, reliable, secure products that possess *interoperability* – compatibility with competitive products. Second is access to distribution. Sales in this industry are built on relationships, as there are a handful of companies in each country that are able to buy this equipment on a large scale. Incumbents such as CISCO have the contacts, credibility and rapport with technology buyers, making it very hard for a new entrant to fund itself a niche in the market. Switching costs are also high: once a customer has a dependable, scalable, interoperable network infrastructure in place, she's not likely to change suppliers without great provocation. In addition, because networks by nature must be interoperable, these companies have to manage open and complicated relationships. That in itself is a demanding balancing act; but in addition, they face the complication of possible competition both from those customers to whom they license technology and suppliers from whom they buy technology.

These characteristics would seem to imply that the profitability of the network equipment sector should be reasonably high. This was true until the end of 2000 when the telecom crash ravaged the industry.

Network Equipment: Strategic Approaches

In the core market, companies are able to achieve competitive advantages by leveraging new technologies and promising better quality and service. Although the field consists of only a few key players, the rivalry among them is intense. Leads are short-lived as competitors make similar or better levels of commitment to their customers. To achieve customer lock-in, companies are focusing on providing quality, customized, interoperable, scalable, end-to-end solutions with exceptional customer service, including in some instances managing the network for the client in what could be called a complete outsourcing contract.

The key players have responded to new market threats by partnering with the operators, not only effectively diminishing competitive nuisance and reducing customer options, but also enabling the entrenched companies to strengthen their offers of end-to-end solutions, making it increasingly difficult for single-component challengers to break in.

Network Equipment: Industry Structure and Trends

Competition is intensifying as many of the smaller telecoms are failing. The remaining players jostle for disenfranchised customers as the customer base is forced to consolidate. However, the overall number of equipment competitors providing niche product solutions is projected to increase; following the same trend, the actual number of larger vendors supplying end-to-end networking solutions will probably decrease due to the rapid pace of acquisitions.

4.4 Web Server Hardware and Software Network Infrastructure

The third segment of the infrastructure stage contains companies that provide both server hardware and server software (applications and OSs). (*Note*: both software applications and OSs will be discussed in detail under Client Software in Chapter 2.) Servers are the machines that do the 'heavy lifting' in designing and operating individual Web sites, of which there existed no fewer than 160 million at the close of 2002. Thus they comprise a fundamental piece of the Internet infrastructure.

Web Server Hardware and Software: Competitive Factors

The Server Market

Until the recession in 2000–2002, The Web server market had been quite profitable, and it can be expected that in an improving economic climate, it will continue to be so. In addition to the hardware, servers have to provide basic software layers, the OS (like any other computer) and the server software itself. The OS layer is primarily a Unix[6]-market, with the Unix servers alone making up a $25 billion industry.[7] The annual growth of the market for Unix servers (12 per cent) is much higher than the rest of the server market (5 per cent). Hardware-wise the leading player in the Unix market is Sun Microsystems, Inc. (Sun) with a 43 per cent market share, followed by Hewlett-Packard (HP), IBM, Dell and Compaq. In addition to the high-end servers based on the Unix OS, competition is intensifying at the lower-end and mid-range server markets due to the growing use of alternative server OSs developed by Microsoft (NT and 2000) and other Linux-based companies.[8] These companies are attempting to close the performance gap between the more powerful Unix-based servers and their own, providing a much more feasible alternative for corporations. Not surprisingly, the server software running on Microsoft's OS is by-and-large the Microsoft Internet Information Server (IIS) in its many varieties while the majority of the 60+ per cent Unix servers run the Apache system.

Demand for Servers

Demand for servers was driven by the new dot.com companies (which as events showed, made it a risky business), continued growth of ISPs, hosting and ASPs, as well as so-called 'old economy' companies that wanted to establish a rewarding presence on the Internet. With the growth of e-business, the need for scalability has increased, because more and more businesses are demanding processing support for a large number of simultaneous users. Despite this growth, however, rivalry is extremely intense at both the high and the low end of the server market. The market leaders for a while outpaced the market by luring business away from second-tier companies. But to continue that growth they turned on one another; price competition is clearly cutting into the once-healthy margins of the market.

Barriers to Entry

Barriers to entry are high due to the huge capital investments necessary for developing both hardware and software. The large global companies competing in this market – HP-Compaq, IBM, Fujitsu–Siemens, Sun and Dell – finance their strategies with massive revenues, tens of millions of which they pour into R&D every year. They offer extensive support and trained technicians, building associations that new entrants cannot easily match. Although it appears on paper that customer-switching costs ought to be low due to the willingness of some customers to switch suppliers virtually overnight, in the end customer relationships are pivotal to staying-power in this business. Once most customers' server systems are installed and running, it requires an exceptionally tempting offer (like those described below) to persuade them to change.

Web Server Hardware and Software: Strategic Approaches

Competitors have been bending over backwards to win Unix business.

They have discounted heavily, sometimes up to 50 per cent off the list price They have been throwing in free software services, giving hardware to young Internet companies and letting them pay later, once they started to generate revenues. They were also buying equity in customers' companies as a way to lock-down deals and ensure future loyalty. Companies were also battling aggressively to acquire customers whose reputation they could leverage to attract additional businesses.

In 1999, Sun decided to become more open by releasing the source code of its own version of the Unix operating system (named Solaris) to the general public as a way of improving its standing among open source developers. The program provides free access to the code for R&D purposes. Programmers can experiment with and improve it, but have to pay licensing fees to Sun if the code is used in commercial applications.

Predictably, Microsoft has aggressively been going after the server market. The company has teamed with Intel (calling again on the so-called 'Wintel alliance'[9]) to address high-end server demand by developing more heavy-duty products. It is growing its server OS market share and increasing its revenue per company server (as it did earlier with the desktop) with its Windows 2000 OS and a series of server applications (an important complement to the OS). It has extensive procedures in place to provide sales and product information and training and support programs. Microsoft has arrangements with over 22,000 solution-provider firms worldwide, and formed a joint venture company with Accenture (formerly Andersen Consulting) in which Accenture will train more than one-third of its workforce, or roughly 25,000 of its consultants, on the Microsoft enterprise platform (Windows 2000, Microsoft IIS Suite and related products).[10] For those without deep pockets and/or a healthy revenue stream, the strategic implication of this turn of events is that it rarely pays to go up against Microsoft.

Web Server Hardware and Software: Industry Structure and Trends

The shift from the desktop computer to the network and to Web services means that Web servers, OSs, and applications must be interoperable, regardless of whether they are proprietary or open standard. While this transformation seems to play directly to Sun's strengths, Microsoft and a range of large hardware vendors are demonstrating that the level of competition will continue to grow for Sun and for anyone else in the vicinity. As we saw in Chapter 2, Microsoft dominates the client browser market, providing it with an extraordinary leverage when selling application servers that need special connectivity to the desktop. Microsoft can guarantee seamless connection from end to end of the internet, a claim that is hard to make by companies that rely on open standards that often have to be agreed upon by committees in which Microsoft is also present. In such an intense environment, customers leverage suppliers' offers against one another to carve out generous terms, greatly reducing server firms' profitability.

Notes

1. Gildertech.com © 2000 Gilder Technology Group.
2. The evolution of France Telecom's average per year stock price in the New York Stock Exchange (NYSE) was: 1998, $61.56; 1999, $90.31; 2000, $128.9; 2001, 51.14; 2002, 18.59. The stock price of KPN in the same market evolved with the same trend: 1998, $47.02; 1999, $52.02; 2000, $58.25; 2001, $7.54; 2002, $5.2.
3. S. Mehta, 'It's Only when you look Closely at the Inner Workings of this $270 billion Industry that you begin to understand why Telecom Crashed', *Fortune*, November 27 2000, pp.124–30.
4. *Transmission Control Protocol/Internet Protocol (TCP/IP)* is a set of rules that determines how computers communicate with one another over the Internet. Any computer or system connected to the Internet runs TCP/IP.
5. In a stock split, the number of shares outstanding increases and the share price becomes a fraction of the pre-split price. In Cisco System there were two splits, both type 2 x 1, between January 1999 and March 2000.
6. *UNIX* refers to a family of OSs based on a system developed in AT&T's Bell Laboratories in the early 1970s. The OS was released from Bell labs as free software, and many different versions were developed by universities, research institutions, government bodies and computer companies. Today the many competing versions of UNIX conform to a single UNIX specification to help standardize the OS. UNIX OSs form a key part of the Internet infrastructure, empowering most of the servers that host Web sites and information for Internet users.
7. L. DiCarlo, 'IBM's Server Story', *Forbes.com*, May 11 2000. For up-to-date statistics on the market share of operating systems and server software see http://www.netcraft.com/survey.
8. Linux is a free, open source OS that runs on a variety of hardware, and is a version of the UNIX OS described above. As an open source software, Linux can be downloaded, tested, used, altered and copied as often as users like, and for free. Companies and developers can still charge money for the OS as long as the source code remains publicly available.
9. The term 'Wintel' refers (sometimes sarcastically) to the alliance between Microsoft and Intel. For years Microsoft has devised Windows programs that run only on Intel's microprocessor architecture, an intimate relationship that has accounted for nearly 80 per cent of the PC market. This duopoly has been viewed by many as unhealthy for the overall computer industry.
10. Deutsche Bank, Alex.Brown, 'Microsoft Corporation: Moving Beyond the Desktop', November 3 2000. http://www.alexbrown.db.com

CHAPTER 5

Hosting

The hosting stage (3) (see Figure 1.3, p. 11) consists of two types of companies – Web hosting firms and application service providers (ASPs). The former provide outsourcing services for corporate customers to manage their hardware, software (creating, storing, and managing data and applications, especially for the World Wide Web) and Internet-access needs. ASPs rent software applications over the Internet to businesses, providing customers with remote access to software applications in exchange for a per-usage or monthly fee.

5.1 Web Hosting and ASPs: Competitive Factors

Hosting companies continue to expect a strong outlook for profits. Margins have been reasonable because growing demand has enabled hosting companies to maintain prices while at the same time benefiting from declining costs of raw materials, bandwidth and storage capacity. Demand for their services is expected to remain strong, as in the face of rapidly changing technology most firms do not possess the internal technical expertise or resources to host their Web sites internally and resort to external help to provide these services.

Hosting Services

Rivalry is intense and consolidation is inevitable. Large and deep-pocketed telecoms, computer giants, high-end consulting firms, expansion-minded Internet service providers and aggressive newcomers are all pursuing essentially the same business, with little differentiation among the services offered. Switching costs have remained high, because once customers select a host, they are reluctant to rework their Web sites in favour of another host with a slightly different deal. But that is likely to change as newcomers compete aggressively to fill their hosting storage capacity. Both established and new companies have been constructing massive host sites, a phenomenon which, to the industry's possible detriment, threatens to create overcapacity. Of course, these companies expect that ever-higher demand will soak up all this new capacity, especially when increasing numbers of 'old economy' businesses decide to include an online presence. A major reason why firms are motivated to provide such extensive hosting services, and why the market is so appealing to telecom carriers, is that customers that buy hosting services often also buy network services.

Barriers to Entry

Barriers to entry are high, since hosting centres are extremely capital-intensive. Large players have quickly built hosting centres while incumbent telecoms capitalized on networks they already had in place. Moving early to grab market share is important because Web hosting is a game of scale. The closer a centre is to a user, the shorter the distance data must travel, avoiding bottlenecks and reducing how much long-haul fibre a host company needs. Web hosts that do not own their own network to connect their host centres are at a disadvantage and must pay network operators to haul their Web traffic. A small number of Internet backbone providers who do own their own network such as WorldCom (a viable company that, probably acquired, will continue to exist even in the wake of the current accounting fiasco) and AT&T benefit by using peer arrangements to exchange traffic free of charge and being able to guarantee rapid delivery for their customers.

The Shift to Networked Computers

For ASPs, the key competitive condition is the shift from the desktop computer to the networked computer, greatly facilitating the on-line renting of software. ASPs can develop their own software or simply purchase it from other developers. While only a small fraction of businesses rent their software, revenue from hosted applications could reach over $25 billion by 2004, based on estimates from a Dataquest study.[1] And, perhaps contra-intuitively, most of that revenue is expected to come from smaller and medium-sized companies, who cannot afford to purchase powerful software.

5.2 Web Hosting and ASPs: Strategic Approaches

Hosting vendors are expected to increasingly own and operate network and data centre facilities, offer end-to-end global outsourcing services and develop specialized offerings to meet the complex requirements of niche and vertical markets. The building up of end-to-end services is coming mainly from the hardware and telecoms industries. IBM is teaming up in strategic alliances to build and operate as many as 49 advanced e-business hosting data centres in addition to the133 data centres it already has world-wide, while AT&T intended to have 44 centres with more than 3 million square feet open worldwide by the end of 2002.[2]

But that is far from the end of the matter. As ASPs gain momentum and display profitability promise, the world's leading software company has jumped on board. Microsoft has signalled that it is intent on transforming itself from being 'merely' a traditional software business that sells pack-aged software to a broad software service company that also rents software. In late 1999, Microsoft began to endorse online application hosting by actively encouraging companies to rent Microsoft's sophisticated business applications online as long as they obtained certification from Microsoft.

5.3 Web Hosting and ASPs: Industry Structure and Trends

As the Internet expands its reach, ASPs will continue to grow apace, espe-cially in view of Microsoft's active involvement. But convergence is taking place in this stage of the value chain as well, with hosting companies and telecoms combining forces. This consolidation is expected to continue, ulti-mately leaving a handful of global players to dominate the market once demand settles down. With their existing brand, infrastructure and installed base telecoms are particularly well positioned to expand into this stage.

5.4 The Role of Storage

Without the ability to store and vast amounts of information and disgorge selected portions of it on command, a computer's contribution to commerce and society would be much diminished. Indeed, we would not be talking of an exponential revolution, but merely an arithmetic increment somewhat beyond the levels Guttenberg stimulated with the invention of moveable type.

What is 'Storage'?

When we speak of 'storage', we refer to the contents of a device in which we leave codified electronic data to be integrally retrieved later on. In computers, the storage sector has markedly advanced from the days in the 1950s when perforated cards were used to introduce (usually statistical) data into a computer, when stout magnetic tapes were used to retain and later disgorge that data, and when keypunch operators were skilled labour in short supply. Storage capacity has advanced by such orders of magnitude that the upsurge in performance has been rivalled only by its rapidly declining costs.

Magnetic Hard Drive

The first significant step toward large-scale data storage was taken in 1956, when IBM announced a magnetic hard drive specifically intended for data storage.[3] The Random Access Method of Accounting and Control 305 was a disk array consisting of 50 24-inch-in-diameter platters stacked on a central rotating spindle inside a cabinet. Data was stored on concentric tracks on both sides of each platter. A single pair of read/write heads could randomly access any track on any of the 50 disks. ENIAC's capacity – 5 megabytes – was impressive for its time. Following the steep slope of computing power's ascent, storage capacity has been increasing about 60 per cent annually. In 2002, even a relatively simple handheld music player (Apple's iPod MP3 player) contained a built-in 10GB hard drive.

Surface Density

Surface density is another concept that affects storage capacity. It refers to how tightly data bits (short for 'binary digits', the basic units of information) can be packed onto a given surface. By 2002, the ability to squeeze information onto the face of a computer's rotating hard drive had reached around 6GB per square inch. But under technology harnessing the quantum properties of matter, it is expected that that number will soon reach 56GB. Some hold that the theoretical limit applying quantum technology to magnetic storage is about 300GB per square inch (Figure 5.1).

Other Physical Advances

While more and more data is being packed into storage devices, the cost of a megabyte's worth of storage has been plummetting. In 1977, a

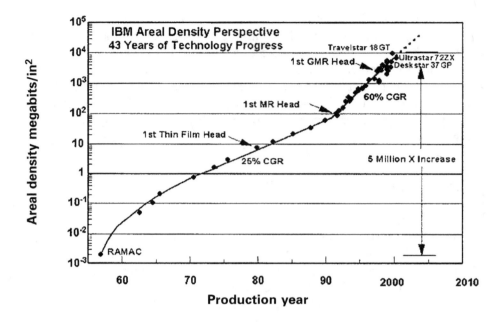

Figure 5.1   Surface density and storage capacity

Source: IBM.

megabyte's worth of storage cost about $1000, that same amount today has dropped to only a few cents (see Figure 5.1). And the average office machine disk has evolved from a few hundred *kilo*bytes of capacity at the beginning of the 1980s to 100*GB* in 2002. In addition to cost benefits, advances in storage technology have brought about two major physical benefits: the dimensions of storage surfaces have markedly diminished, and stability and durability have greatly improved. The actual weight of, say, two Beethoven symphonies on an optically-scanned compact disc or *Gone with the Wind* on a digital versatile disk (DVD) is less than a gram; further, that disc can be left out in the rain or passed through x-rays and magnetic fields without injury to the data it holds. (You cannot, however, scour it with sandpaper, as the microscopic pits that store the 1s and 0s of its digitized data will be ruined.)

Applications

Popular applications such as games and movies are obvious stimulants to innovation in data-storage technology. But there are many other present and future applications that demand continued research and development in this field.

- **E-commerce:** One of the most important applications is *on-line shopping*, which undoubtedly will utilize video and audio content more profusely. A shopper for an automobile, say, will select files that produce 3-D pictures of a chosen vehicle, demonstrating it in full motion and from all angles to produce a 'feel' for its looks and driving characteristics. Similarly, mall-goers will elect to stay at home and browse shopping centres from their desktops, digitally 'handling' the virtual merchandise by choosing objects to examine and rotating them through all visual axes on their computer screens. In this way, furniture shoppers, for instance, will decorate rooms in their homes by viewing virtual furnishings set in place in 3-D perspective.
- **Picture/sound libraries:** With the advent of digital cameras and camcorders with audio/digital (A/V) functions, users can employ a capacious hard disk as a catalogued library; 1,000 still images with standard resolution can be 'shelved' in 1GB of storage. Good-quality video such as produced by the Video Home System (VHS) or the Moving Picture Experts Group – Level 1 (MPEG-1) need 1.5 megabytes per second (mbps) (5.5GB per hour). High-definition television (HDTV) and MPEG-2 require 10GB per hour. Much of this is already enabled by single hard drives. Sometime in the near future, *transferable* hard drives for PCs will make a full-fledged home or office A/V library a reality.
- **A/V email:** Unlike today's primarily text-only e-mail, soon most e-mail messages will support full-motion video (at 30 frames per second, rather than freeze-frame) along with audio content. Many of these large-capacity messages will be downloaded to the hard disk for later reference. Depending on the length of the file and of the degree of compression, the storage requirements of a typical piece of A/V email will be between 2MB and 15MB.
- **Teaching:** As more universities and professional educational institutions offer online education programs, multimedia applications will undoubtedly expand. Multimedia clips will be included into material sent to students to be downloaded for customized study and replied to inter-actively (as, for example, an exam). Busy doctors will be able to keep up to date – a problem in the practice of medicine – by downloading and reviewing the details of the latest treatments and procedures in colour video during less hectic hours.

Notes

1. Dataquest, 'ASP Market to be Worth USD 25.3 Billion by 2004', August 11 2000.
2. B. Wallace, 'Web Hosting Heats Up', *Informationweek*, September 18 2000, pp. 22–5.
3. It could be considered that, according to the magazine *Nature*, 'The first computer to store a changeable user program in electronic memory and process it at electronic speed' was a machine developed at Britain's University of Manchester in 1948. The apparatus, called the Small-Scale Experimental Machine and nicknamed 'The Baby', was designed and built by Frederic Calland Williams and Tom Kilburn. The Baby kept only 1,024 bits in its main store (*Nature*, 162 (487), 1948).

CHAPTER 6

User Hardware

Stage (4) of the value chain deals (see Figure 1.3, p. 11) with the firms who build the computer devices that end-users employ to access the Internet. The devices run software that allows them to make requests of servers over an Internet protocol (IP) network. Hardware devices have become much more diverse than just desktop and portable PCs. Fresh interactive implementations appear regularly, and now also include personal digital assistants (PDAs, also called handhelds or palmtops), third-generation mobile phones, satellite phones, televisions, entertainment/game consoles and more.

6.1 User Hardware: Competitive Factors

At the turn of the new millennium, hardware manufacturers suddenly found it rather difficult to make money. Competition had become intense and margins thin. Prices fell into a swoon. Life cycles of PCs, PDAs and even mobile phones contracted. These once-chic and relatively dear devices had rapidly become everyday household items.

In this stage, commodity status reduces customer switching costs as well as barriers to entry – 'no-name' manufacturers, often using identical internal parts, can easily compete with companies that have over the years had at some cost built up brand identification, distribution channels and sales forces. However, the value of this hard-won structure shrinks markedly as shoppers become comfortable with buying such appliances online, receiving much the same guarantees and quality of support and service from the 'no-names' as from the known-names. Indeed, sometimes it is the latter – such as the financially squeezed Gateway Computer in the USA – which worried shoppers more.

PC History

The history of the PC Industry may be relevant at this point. Apple pioneered the first usable personal computing devices in the mid-1970s, but

IBM was the company that brought PCs into the mainstream when it entered the industry in 1981. IBM's brand name and product quality helped it to capture the lion's share of the market, including almost 70 per cent of the *Fortune 100*. Most IT specialists chose IBM instead of cheaper clones due to fears about quality, compatibility, reliability and service. In the late 1980s, IBM saw its dominance eroding as buyers increasingly viewed PCs as commodities. IBM tried to develop a more proprietary PC but failed because the open combination of Windows (developed in early 1990s) and Intel microprocessors (see Exhibit 6.1), called 'Wintel', got established as the dominant design or standard in the PC industry, replacing the 'IBM-compatible' device. Throughout the 1990s, thousands of manufacturers, ranging from Compaq and Dell to 'no-name' clones, built PCs around the standard building blocks from Microsoft and Intel. By 2002, there were over 400 million PCs installed around the world and the PC industry was a $220 billion global industry.

Exhibit 6.1 Intel and the Microprocessor

Microprocessor

A microprocessor is a silicon chip that contains a CPU (Central Processing Unit). The CPU is a circuit that interprets and executes programs by processing a list of machine instructions which perform arithmetic and logical operations, and decodes and executes instructions. In micro-computers the entire CPU is on a single chip. In the world of PCs, the terms 'microprocessor' and 'CPU' are used interchangeably. There is a microprocessor at the heart of all PCs and workstations. Microprocessors control the logic of almost all digital devices, from clock radios to fuel-injection systems for automobiles.

The History of Intel

Intel, which is a contraction of 'integrated electronics', is the most famous microprocessor company in the global market. It was founded by Robert Noyce and Gordon Moore in 1968. Their main goal was to make semiconductor memory more practical.

Intel initially provided computer memory chips such as DRAMs (1970) and EPROMs (1971). Later, microprocessors appeared. Table 6.1 shows the evolution of Intel Microprocessors.

Table 6.1 Intel microprocessors, 1972–2001

1972: 4004 Microprocessor	The 4004 was Intel's first microproces sor. The chip was capable of executing 60,000 operations in 1 second
1972: 8008 Microprocessor	The 8008 was twice as powerful as the 4004
1974: 8080 Microprocessor	The 8080 became the brain of the first personal computer, the Altair
1978: 8086–8088 Microprocessor	A pivotal sale to IBM's new personal computer division made the 8088 the brain of IBM's new hit product, the IBM PC
1982: 286 Microprocessor	The 286, also known as the 80286, was the first Intel processor that could run all the software written for its predecessor
1985: Intel386™ Microprocessor	The Intel386™ microprocessor featured 275,000 transistors, more than 100 times as many as the original 4004
1989: Intel486™ DX CPU	The Intel486™ processor was approximately 50 time faster than the original 4004 processor
1993: Pentium®Processor	Pentium was five times as fast as the 486 and was capable of 90 million instructions per second (MIPS)
1995: Pentium® Pro Processor 1997: Pentium® II Processor 1998: Pentium® II Xeon™ Processor 1999: Celeron® Processor 1999: Pentium® III Processor	The processor incorporated 9.5 million transistors
1999: Pentium® III Xeon™ Processor 2000: Pentium® 4 Processor	The processor debuted with 42 million transistors and circuit lines of 0.18 microns Intel's first microprocessor, the 4004, ran at 108 kilohertz (108,000 hertz), compared to the Pentium® 4 processor's initial speed of 1.5 gigahertz (1.5 billion hertz)
2001: Intel® Xeon™ Processor 2001: Itanium™ Processor	The Itanium™ processor was the first in a family of 64-bit products from Intel

In mid-2001 Intel announced a major deal with Compaq under which Compaq would use the Itanium architecture in all of its high-end servers by 2004; meanwhile Intel gained a licence to Compaq's Alpha chip technology. Intel later absorbed some Compaq chip design teams, as well as Itanium-related engineering groups from Hewlett-Packard, even before those two companies announced their gigantic merger.

Later in 2001 the company announced that it would phase out its consumer electronics operations (Internet appliances and toys). In a move widely seen as an indication of the company's succession plan for top management, Intel promoted Executive Vice President Paul Otellini to president and Chief Operating Officer early in 2002. Also in 2002, the company announced that it would close its Intel Online Services unit, which provided Web hosting services, by mid-2003.

Today Intel's business can be broken down into four categories: client platforms, server platforms, networking and solutions and services. Client platforms are the equipment necessary to connect to the Net. Server platforms feed out all the information. In the last 3–4 years, Intel has expanded its networking operations, the back-end technologies that enable the network on which the Internet runs. Finally, Intel is also making inroads into the business solutions field, offering start-ups new products such as e-commerce, back-end systems and online services.

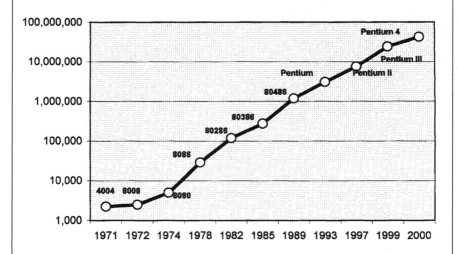

Figure 6.1   The evolution of Intel's microprocessors,
1970–2000

Source: Intel.

Moore's Law

Gordon E. Moore is currently Chairman Emeritus of the Intel Corporation and co-founded Intel in 1968. Moore is widely known for 'Moore's Law', in which he predicted that the storage capacity of computer microprocessor chips would double approximately every 18 months (Figure 6.1). While originally intended as a rule of thumb in 1965, it has become the guiding principle for the industry to deliver ever-more-powerful semiconductor chips.

Hardware Devices

End-user hardware now combines elements of the consumer-goods mass market (design, branding, distribution, support) along with the innovation that has always been expected of it. Despite some ability to differentiate themselves through technological novelty, companies can rarely defend themselves via patents, and most hardware devices are subject to virtually instant replication by both offshore and domestic manufacturers. As a result, stock hardware entities such as the desktop computer or even PDAs tend to become low-cost commodity items, existing in a crowded market filled largely by the familiar handful of multinational manufacturers doggedly turning out goods through high-volume, low-cost production modes.

Complementary Elements

Given the diversity of the IP environment, the hardware device can no longer be considered in isolation. Key competitive factors have shifted from a focus on hardware features to complementary elements such as quality of connection to the Internet and range of software applications. Because the PC has a longer history than the other devices, it has far more software applications than the others do. In addition, PCs tend to be more powerful than smaller, more mobile devices; therefore, they have traditionally offered greater performance. However, hardware size is becoming less of a factor as microprocessor technology is enabling small, wireless devices to carry out more robust computer functions.

Network externalities and positive feedback are playing important roles as expectations intensify for non-PC devices. Programmers anticipating large and growing markets are dedicating their efforts to creating applications for these new devices, which makes them even more attractive to consumers. This positive feedback flows back and forth in building non-PC markets. Because of the importance of complements, non-PC device manufacturers are working aggressively to set standards and form alliances with complementors,[1] such as the Symbian alliance to produce smartphone software, which includes manufacturers as Nokia, Motorola, Sony-Ericsson, Siemens, Panasonic and Samsung, accounting for almost 80 per cent of all handsets sold in 2002.

Non-PC Internet devices (also called Web appliances) are expected to make up a large share of the Internet device market, far outstripping the arithmetically well-established PC in growth. As of 2000, only 2 per cent of Internet users in the USA and 6 per cent of Internet users worldwide accessed the Internet with a non-PC device. However, recent predictions project worldwide usage to soar to over 70 per cent by 2005 for at least some of users' online activities.[2] Leading this growth is the increased availability of Internet-ready mobile phones, the third generation of which boast larger screens. Since non-PC devices are widely seen as complementary to the PC market, the battle is not between the PC and wireless devices as much as it is among the wireless devices themselves, especially between phones and PDAs. Indeed, in the not-too-distant future it is very likely that each will incorporate the functions of the other, although at present the two wireless devices are divided technically by a gap in interoperability.[3]

6.2 User Hardware: Strategic Approaches

Low-Cost Production

As hardware trends towards commodity status, one proven strategy is to be the lowest-cost producer. Dell Computer (see Exhibit 6.2), the most profitable PC manufacture in the world (as of 2002), has successfully built a cost-leadership approach through its direct selling model. In addition to price, the Dell model offers built-to-order customization as well as online information and responsive customer service. Finally, Dell focuses on corporate customers where switching costs tend to be higher. However, the company's main competitive advantages are increasingly being imitated, at least in part (by Gateway, Compaq and Sony, among others).

Exhibit 6.2 Dell Computer

Michael Dell founded Dell Computer with the revolutionary idea of selling customized computers directly to the customer. When he began his studies at the University of Texas in 1983 he was interested in the entrepreneurial world. He spent his evenings and weekends re-formatting hard disks for IBM-compatible PCs. In 1984, Dell earned about $80,000 a month, enough to persuade him to drop out of college to attend to his business. The company was plagued by management changes during the mid-1980s. In 1988, the company started selling to larger customers, including government agencies. Dell went public that year.

By 1990, sales had grown even further to over $500 million. Dell's sales were to large corporate accounts, medium- and small-businesses, federal and state governments sized and educational institutions. The end-consumer market was only a small proportion.

In 1996, Dell began conducting business through its Internet site and pioneered the Dell Direct Model ('made-to-order' system) for computers. Online order history and business forecasts were used as the basis for purchasing and manufacturing activities. Dell never assembled a computer system until it had received an online order, so every system the company made already had a waiting customer. Although the PC market was price-sensitive, Dell Direct Model was very efficient, with high speed, low distribution costs, sales force efficiency and direct customer relationship. Today, Dell is the cost leader producer in the segments it competes in.

Once customers registered and configured the system that best suited their needs, they had the option of purchasing a computer using a credit card. The online order was received by an order processor which classified the order into the market segment.

Today more than 90 per cent of Dell's purchases are handled online: suppliers forecast their activities using an Internet portal that lets them view Dell's requirements and changes based on market place activity. As Dell factories receive orders and schedule assemblies, a 'pull' signal to the supplier triggers the shipment of only the materials required to build current orders, and suppliers deliver the materials directly to the appropriate Dell assembly lines. This online system helps suppliers meet Dell's delivery requirements.

Dell schedules every line in every factory around the world every two hours, and it brings into the factories only 2 hours' worth of materials. Dell has decreased its cycle time and has reduced warehouse space, which has been replaced by more manufacturing lines.

Dell has recently tried to change its strategy from the PC market to the computer systems and servers market, where IBM and Sun Microsystems are the main competitors. Despite the fact that Dell has $31.2 billion in revenues and a 28 per cent share of the US PC market, the high-margin business of servers, printers and other equipment is more profitable than the tight margins of the low-end PC business. Table 6.2 shows Dell's financial statistics from 1999 to 2002.

Table 6.2 Dell detailed annual financials, 1999–2002

Income Statement

All amounts in million dollars except per share amounts	Jan 02	Jan 01	Jan 00	Jan 99
Revenue	31.168	31.888	25.265	18.243
Cost of Goods Sold	25.422	25.205	19.891	14.034
Gross Profit	5.746	6.683	5.374	4.209
Gross Profit Margin (%)	18	21	21	23
SG&A Expense	3.236	3.675	2.761	2.060
Depreciation & Amortization	239	240	156	103
Operating Income	2.271	2.768	2.457	2.046
Operating Margin (%)	7	9	10	11
Nonoperating Income	−58	531	188	38
Nonoperating Expenses	0	0	0	0
Income Before Taxes	1.731	3.194	2.451	2.084
Income Taxes	485	958	785	624
Net Income After Taxes	1.246	2.236	1.666	1.460
Continuing Operations	1.246	2.236	1.666	1.460
Discontinued Operations	0	0	0	0
Total Operations	1.246	2.236	1.666	1.460
Total Net Income	1.246	2.177	1.666	1.460
Net Profit Margin (%)	4	7	7	8

Source: Hoover, 2003.

R&D Investment

Another strategy is to invest in R&D to stimulate demand for new products. Differentiation is difficult to sustain in hardware because of the computer's commodity status. However, since non-PC markets are still rapidly developing and tend to have lower price points, differentiation is more achievable and repeat purchases more frequent. For example, demand for new mobile phones is increasing as mobile phone applications expand past simple email and into more complex Internet areas such as streaming video. Thus mobile phone users tend to replace their phones more often than they replace their PCs.

Service Contracts

Hardware prices have been driven down so forcefully that in some cases it is more profitable to support and service them for a fee. To set the stage for such fees, hardware is sometimes given away free to consumers who will commit to a multi-year service contract. Mobile phone and cable TV companies adopted this strategy in the past to get customers to buy their service. This trend will further commoditize PCs to a point where the brand name on the box may be less important than the name of the company providing the Internet access.

PDAs

One may claim that the same trend towards commodity status that affected the PC will eventually affect all other devices which are being developed to connect with the Internet. However, companies playing in these new fields understand the history of PCs and will do everything possible to avoid commoditization, or at least, to delay it as much as they can, and eventually control the key elements either with open or proprietary standards. Developments in PDAs and mobiles could be illustrative here.

The first successful PDA was Palm, an integrated device that incorporated a Palm OS. Its fast success caught the attention of Microsoft, which developed a reduced version of its PC OSs, called Windows CE, and licensed it to manufacturers like Compaq. Today both OSs are being licensed and the Palm OS still holds a 50.2 per cent market share, but Windows CE with 28.3 per cent[4] is growing fast by leveraging on Windows, MS Outlook and MS Office, and the ability to connect with this broadly used software. Table 6.3 presents the market share of PDA shipment forecast by OS, as estimated by Gartner Dataquest (October 2001).

Mobiles

The mobile industry developed rapidly in Europe thanks to the open standard philosophy around GSM. In most countries mobile penetration is

Table 6.3 PDA shipment forecast 2000–5, by OS

OS	Market share (%)					
	2000	2001	2002	2003	2004	2005
PalmOS	66.8	55.8	43.6	38.7	34.0	30.9
Windows CE	17.1	23.0	38.8	42.3	48.5	50.4
Linux	0.0	0.6	2.5	5.0	8.5	11.7
RIM3.8	4.2	3.0	1.8	1.1	0.8	
Symbian EPOC	4.2	1.8	0.2	0.1	0.1	0.0
Others	8.1	14.7	12.0	12.1	7.9	6.2
Total	100.0	100.0	100.0	100.0	100.0	100.0

Note: Some columns do not add to the totals shown because of rounding.
Source: Gartner Dataquest, October 2001.

larger than PCs, making the handset and attractive device for Internet connection. Even with all delays and difficulties associated with the deployment of G3 'third-generation' mobile networks, there is still a possibility that mobile phones may be taking over from PCs as the focus of the entire technology industry. Handset manufacturers will do everything to avoid reaching commodity status.

Two key aspect can be highlighted in their strategy. On the one hand, they all have an interest that some key features of the telephone are incorporated in the hardware so that the performance of the handset is not completely dependent only on the software it may incorporate. In this way, some degree of differentiation is still possible even with common software.

Software is key for the mobile phones to connect with the Internet. Manufacturers are aware of this and they are trying to keep it as proprietary as possible, integrated in the handset and based in open standards as they did with the development of GSM; this is what they tried to do with the Symbian alliance. However, as usual, Microsoft is trying to develop its own operating system and launched SPV. With its Windows-based software, the SPV is, in effect, a PC crammed into the casing of a mobile phone, complete with versions of Microsoft's web browser, e-mail and media-playback software. As Microsoft did in the PC, here it is trying to position itself to monopolize the operative system as hardware becomes a commodity.

While the manufacturers defend themselves with Symbian (see Exhibit 6.3), Microsoft tries to get round them by going directly to their customers: the mobile-network operators which buy handsets in bulk and sell them to their subscribers. This is possible because 26 per cent of the handsets in 2002 were built by contract manufacturers.[5] The SPV phone will then be a joint venture of Microsoft, HTC (a Taiwan contract manufacturer) and Orange (a European mobile-network operator, very strong in the UK, and very innovative in its marketing approach. Orange gets a customized phone and Microsoft's OS reaches the market place. Several Symbian-powered handsets have already come to market; the latest is the Nokia 7650 (see Figure 6.2 in Exhibit 6.3), with a built-in camera and colour screen. Other phones will reach the market with the same platform and very different features. It is too early to conclude if Microsoft is attempt to by-pass the handset makers will fail or not, or how the whole industry will develop. The fight is on, but we have the lessons hard learned from the PC industry, and the mistakes to avoid.

Exhibit 6.3 Nokia and Sony-Ericsson

Nokia

Nokia is one of the world's leading players in wireless communications. It has always been involved in the development of new technologies, products and systems for mobile communications. Recent examples include the co-development with Symbian of the OS for future terminals.

Nokia has mobile services as a key part of its strategy. As demand for wireless access to an increasing range of services accelerates, Nokia plans to lead the development of the higher-capacity networks and systems required to make wireless content more accessible and rewarding to the end user. Nokia's strategy focuses on driving open mobile architecture, enabling a non-fragmented global mobile services market; being the preferred provider of solutions for mobile communications; creating personalized communication technology; and expanding its business and market position on a global basis (Figure 6.2).[1]

Sony-Ericsson

Sony Ericsson Mobile Communication AB is the mobile-telephony joint venture created by telecoms leader Ericsson and consumer electronics powerhouse Sony Corporation in 2001. It was projected to become the world's leading mobile phone maker by 2006. Sony-Ericsson is responsible for product research, design and development, as well as marketing sales, distribution and customer services of mobile multimedia consumer products.

Figure 6.2   The Nokia 3650

Source: Nokia.

Although Sony-Ericsson announced its first joint products in March 2002, it has disappointed its parent companies, losing market share to competitors like Nokia, Motorola, Samsung and Siemens. In 2002, Sony Ericsson posted an after-tax loss of 241 million euros, or about $260 million (Figure 6.3).

Symbian

The leading-edge technology companies Nokia, Ericson, Motorola and Psion established Symbian as a private independent company in June 1998. Panasonic joined Symbian as shareholder in 1999 and Siemens and Sony-Ericsson in 2002 (see Figure 6.4). Symbian is a software licensing company that will supply the open standard OS (Symbian OS) for data-enabled mobile phones with mobile networks, content

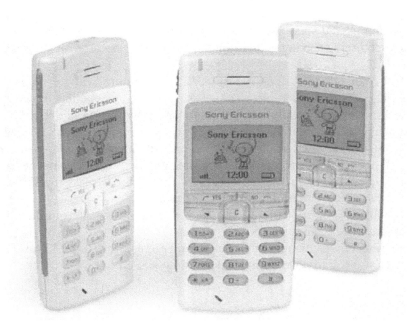

Figure 6.3 The Sony-Ericsson T-2102

Source: Sony-Ericsson.

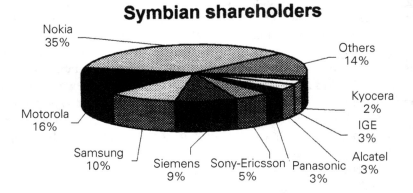

Figure 6.4 Symbian

Source: Symbian, 2003.

applications and services. Symbian OS enables leading mobile phone manufacturers to bring mobile phones to market because of its open standard.

Symbian OS integrates the power of computing with telephony, bringing advanced data services – using voice, messaging and on-board processing power – to the mass market. Symbian OS is flexible and scalable enough to be used in the variety of mobile phones needed to meet a wide range of user requirements. Its open standard ensures global network interoperability. Consequently, software developers are able to build applications and services for a global mass-market of advanced, open, programmable, mobile phones.

Symbian is headquartered in the UK and has offices in Japan, Sweden, the UK and the USA.

Symbian helps manufacturers to defend themselves from Microsoft, which is trying to develop its own Windows-based OS.

Note:
1. Source: www.nokia.com.

6.3 User Hardware: Industry Structure and Trends

Squeezes on hardware margins and pressures on profitability have inspired some vendors to concentrate on revenues 'beyond the box'. This tactic involves promoting such services as Internet access, Web hosting, financing, add-on peripherals (such as LCD terminals), maintenance and support to their hardware customers. The goal is to generate a stream of recurring revenue over time rather than just at the initial point of POS sale for the hardware. As a result, the hardware industry has converged with several other stages along the network in its quest to uncover profitable markets to address. Approximately one-quarter of loss-troubled Gateway's revenues come from non-PC-manufacturing sales, and the company says that eventually it intends to derive all its profits from offerings such as Internet access, with PC sales effectively serving as a loss leader. Dell Computer CEO Michael Dell has noted that the Internet is pulling companies like Dell in two directions – toward the core, where servers and huge storage systems reside, and toward the edge, comprising PCs, mobile phones and handheld devices.[6]

Dell is now developing a new enterprise strategy.[7] Consistent with the philosophy to move towards the core, Dell believe he can attack the expensive mid-range servers produced by IBM, HP and Sun by clusters of cheap low-end servers (where Dell is growing very fast and is already in second position) acting as a single high-performing unit running on either Windows or Linux operating systems. The idea is to get the same performance or even better with a fraction of the price of expensive big Unix machines. Dell has also announced plans to start making printers and hand-held computers and could move to other complementary devices. The low cost strategy will be present in all segments and accelerate the trend towards commodity status. This pressure will stimulate strategies to 'go beyond the box' into other, more profitable and structurally attractive stages in the value chain.

Notes

1. A product is considered a *complement* to your product if by using it the user values your product more. A product is considered a *competitive* product if by using it the user values your product less.
2. eTForecasts, 'By 2005 55 per cent of US Internet Users Will Use Web Appliances Internet Access via Multiple Devices Is Growing', June 12 2000.
3. D. Lillington, 'Mobile But Without Direction', *Wired.com*, September 21 2000.
4. *Source*: Gartner Dataquest, October 2002.
5. Data coming from 'Special Report: Nokia and Microsoft', *The Economist*, November 23 2002.
6. D. Lyons, 'Michael Dell's Second Act', *Forbes.com*, April 17 2000.
7. 'Dell's High Risk, High-End Dream', *Financial Times*, December 19 2002.

CHAPTER 7

Proprietary Content Providers: Aggregators (Portals)

Although stages (1) and (2) in Figure 1.3 share much in common, it will be useful to distinguish their different approaches to serving customers by treating each separately. In section 7.1 we will look at content providers (1) and in section 7.2 we will focus on portals (2).

7.1 Content Products and Services

Stage (1) – products and services – includes companies along a very varied spectrum that can be grouped, based on the purpose of the offering, into five broad categories: entertainment, business information, learning, transactions and personal communications. A full study of all these is beyond the scope of this book,[1] here we will study two segments that we feel illustrate the most relevant aspects of this stage: (a) content providers and (b) e-commerce companies. *Content providers* include individuals or companies that develop and/or distribute goods that can be represented and delivered in digitized form, such as text, data, audio and video. *E-commerce* includes individuals or organizations that trade or facilitate trade over the Internet, such as online stockbrokers and 'e-tailers'.

Proprietary Content Providers

The proprietary content provider segment consists of a large and varied range of information and entertainment companies dealing in digitizable material from text to moving visual images. The focus here is on two types: (a) originators or what the industry calls 'talent' – those who create new content, such as writers or musicians; and (b) packagers – those who traditionally have packaged and continue to package (and now often deliver) content, such as movie producers, newspapers or book publishers. Both differ from portals by exclusively owning the rights to content for a given period of time and/or geographic area. (A portal may well provide financial services, entertainment, and so forth, but this does not constitute proprietary content *per se*; the portal's added-value that we want to study here comes from the aggregation of others' content into categories that facilitate browsing the Internet or any other network.)

Proprietary Content Providers: Competitive Factors

Cost Structure Digitized goods defy the basic economic law of scarcity – or at least they have established a new basic law: when dealing in binary bits, there never *is* scarcity. Bits don't have shelf lives (unless the power is inadvertently turned off), and digitized goods don't get used up, even if they're sold many times over. Nor is there such a thing as inventory management; a seller 'warehouses' essentially the identical good, since normally there are no capacity limits to the production of additional copies. Thus the seller can continue selling the good virtually ad infinitum at zero incremental costs. Indeed, a unique factor of digitized content is its cost structure: although the up-front production of information often is relatively costly (high fixed costs, normally sunk costs), the subsequent *re*producing of it is not at all costly (marginal costs are virtually nothing). This cost structure dictates yet another law relating to digitized goods: profits increase rapidly – at precisely the same rate that sales increase, there are no variable costs.

IPR Another aspect we must take into account when designing and marketing information-based products is the adequate defence of intellectual property rights (IPR).[2] Management of these rights is also a basic strategic variable. The efficiency of digital technology affects both the costs of reproducing information-based products and the distribution of them (software, for example, can be downloaded free, and, of course, in that delivery mode costs nothing to produce). The drastic reduction in distribution costs permits yet other tactics to increase the installed base of customers. These tactics include simply giving away the product (at least initially), giving away a similar but not identical product, offering a product without service included, supplying a sample or time-terminated product and so on – each of which feeds the accumulation of an installed base.

Digitized information can be delivered globally, instantly and at very low cost. These factors, plus the ease of reproduction and distribution make the protection and management of proprietary IP vital to its owner. That has proved to be a challenge not easily solved. On the negative side, however, is the unfortunate but well-proven fact that the low distribution costs of IP also enable illegal dissemination of digital products. (Looking after its own, however, the same technology has developed relatively effective curbs on illicit activity.) Low reproduction costs make it more important to handle IP stocks efficiently. Liberal policies on software licences or even photo archives, for example, increase the value of the product for customers, but they also reduce the number of billable transactions. For years in the music industry, performers, composers and record producers and even the Recording Industry Association of America repeatedly called on the US courts to help protect their property rights against download-for-free Web entities such as Napster, Kaza, Audiogalaxy and Gnutella. Despite a US court victory in 2001 barring Napster from exchanging music openly over the Net, the music and movie industries have not been able to keep up. Schemes continue to proliferate. In 2002, peer-to-peer Internet free-swapper Audiogalaxy (for one) turned to the older zip compression to disseminate music. Using zip compression, an entire album (not just one song, as with MP3) was downloaded in a single file – together with the album's liner art and notes.

Disintermediation   At the same time, traditional third-party content packagers of original material such as books or music are threatened with loss of business through disintermediation (a process emanating from the efficiency of networked markets, which cuts out middlemen such as retailers and/or wholesalers) as lowering costs and increasing ease of distribution enable the content's creators or wholesalers to bypass the packagers and deliver the goods themselves.

As a result, access by content providers to unique and exclusive content is crucial to achieve differentiation and to avoid being disintermediated. Competition for such content is intense, of course, causing prices – and therefore barriers to entry – to rise. The fact that many commercial Internet sites proffer free information (although sometimes as a vehicle within which to run paid advertising or as a lure to attract paid subscribers) is a result of the competition that has pushed the price to marginal cost – in this case, essentially zero.

Narrowcasting Model   Proliferating information technologies on the Internet have demolished the conventional tradeoff between richness of information and range of its receptivity.[3] Content companies can now convey richness (detail) – instantaneous, detailed, interactive, multimedial, on-demand customized information – to audiences anywhere on the globe (reach). This amalgam – enabled by advances in data transport, larger processing capacity and the development of content protocols such as HTML – allows companies to abandon the old broadcasting model, in which one message is dispatched to a large, undifferentiated audience, to a narrowcasting (or pointcasting) model, in which the content can be tailored to each individual. Thus the Internet has brought about an important shift in bargaining power from the seller to the buyer by allowing the latter to dictate the nature of the content she wants to receive. A subscriber, for example, can personally select from a wealth of proprietary content and services offered by providers such as AOL–Time Warner. (For more information see Exhibit 7.1.)

Proprietary Content Providers: Strategic Approaches

Content Originators   Content originators can harness the Internet to appropriate more of the value they create. Well-known authors such as Stephen King have promoted and sold their own original books online directly to individual readers, thereby disintermediating traditional publishers, wholesalers, and retailers and capturing all the profit for themselves. These approaches have as yet worked only in niche markets such as in market research pay-per-view schemes. In literature such self-publishing initiatives have met with limited commercial success (avid readers still prefer real books, and have been cool to electronic books as well). The prestige, market making and promotion power of large publishers combined with the low penetration of the Internet have offset the appeal of a direct approach to the market. However, in music, sites like MP3.com legally sell recordings submitted directly by the artists themselves, who bypass the customary recording studio sponsorship – and cut of the revenue. The movie studio MGM downloads entire movies from its archives directly to an end-user, coding the downloaded copies to 'self-destruct' in 24 hours. Faced with technological and marketing trends like these, conventional intermediaries saw the writing on the wall, and many have formed exclusive partnerships with, or actually acquired, original-content producers.

Content Packages   The marketability of news-type information on the Internet by content packagers such as newspaper companies depends on their ability to separate their service from the background noise of similar sites: how well they differentiate their material and how attractively they present it. In this sector, traditional packagers of commodity information like stock market commentators, who attract many of their buyers simply through their brand name, won't necessarily enjoy the same distinction when transporting their business online; strong branding in the off-line community doesn't translate automatically into a strong branding position online. For revenue, some publishers operate a subscription service for current material and/or charge fees to research their archives. Even so, the threat remains that they will be forced into a disintermediated status by their own suppliers (such as wire services) who can sell directly to end-customers. More to the point, they face stiff competition from new online intermediaries. On the Internet, the largest financial information providers may not be faithful old steeds like *The Financial Times*, *The Wall Street Journal*, Reuters, or even Bloomberg, but rather interlopers AOL and Yahoo!

Exhibit 7.1 The Reach–Richness Tradeoff

Transmitting information that was not easily codified required a rich, interactive and broad interchange. To maintain the richness of the interaction, however, its range was restricted by systems that had little connectivity – a corporation's private local area network (LAN), for example. By contrast, codified information could be distributed over a wide range and with a high level of connectivity – electronic catalogues, for instance – but at the cost of reducing and rendering ambiguous the richness of the content by generalizing it and not allowing customization.

The effect of the new ICTs is revolutionary, since it is a product of our having improved in both dimensions simultaneously. The ICTs overturn the previously accepted tradeoff by increasing bandwidth, personalization and interactivity while at the same time increasing the connectivity and range of the information.

A vivid illustration of ICTs' shattering the 'either-richness-or-range' shibboleth can be found in the rapid shifts that have taken place within financial institutions such as investment institutions and banking. In a retail stock brokerage industry long dominated by venerable houses like Merrill Lynch, with their reliance on value-added telephone and sent-by-mail services, the young and network-savvy firm of Charles Schwab was able to offer a telephone service (although a stripped-down version) plus lower commission rates, aiming its marketing at the growing number of non-institutional investors who demanded little or no information, but sought a reduction in the otherwise high commission rates accorded to non-professionals. With its 'bare-bones' level of service and concomitant low overheads, Schwab cut huge swathes through the retail brokerage market and became an industry leader in terms of internal costs.

Shortly thereafter, the growth of the Internet delivered a powerful new dimension to the brokerage business: the on-line securities transaction. 'E-trade' gave the public investor professional-level performance that they had never before enjoyed: real-time quotes, immediate executions and super-low commissions (for better or worse, breeding a novel category of investor: the amateur day-trader). At the same time, the Internet also gave Schwab the opportunity to enrich its elementary information service by providing reports, offering on-line analyses and delivering other tit-bits such as corporate news items. That automated package developed into a personalized service capable of competing with those of Merrill Lynch – and Schwab delivered it at a considerably lower price.

The retail brokerage sector was staggered by Schwab's bold incursion. Merrill-Lynch was forced to respond by developing its own electronic system. Among other once-staid banking institutions, a similar but more rapid process took place. Larger traditional banks, such as Spain's Banco Bilbao Vizcaya Argentaria (BBVA) formed alliances with young and able Internet companies to proffer a broad range of cyber-banking services in order to take the market away from slower-to-respond wooden-desk counterparts.

This led to an underlying challenge inherent in the similar transformations many sectors are undergoing: the difficulty of adjusting business models to the opportunities provided by ICTs. There is a tendency to think that all we need to do is to new channels or new ways of gaining access to customers to remain competitive. But in fact we have an opportunity to forge not a reactive but a truly innovative model.

Versioning Strategies   The Internet does enable content providers to boost profits through defined price discrimination strategies. Because the Internet enhances a firm's ability to learn about individual customers, firms can more effectively identify customer groups and offer different prices to the different segments based on their level of demand. If groups are difficult to identify, a versioning strategy can be used. Firms can offer different versions of the same product and customers can self-select the appropriate version based on their needs or level of interest – charging more for earlier releases than for later releases of the same product, or charging more for full access than for limited access, for example.

Proprietary Content Providers: Industry Structure and Trends

Improved ease of access to content through the ever-more diverse and sophisticated platforms that the Internet enables is bringing the client closer to the primary source of the content. As a result, the need for such primary players to acquire and control proprietary content has unleashed one of the immutable forces driving the online value chain, accelerating consolidation in the media sector as packagers rush to acquire content that will differentiate themselves from competitors. This phenomenon is moving traditional content packagers such as Disney, New Corporation, Sony, Vivendi Universal and Viacom to occupy either end of the chain. The big players get still bigger, becoming master of exclusive entertainment and educational material, manager of content distribution and provider of the technology to deliver it – all under the same roof. Smaller content-providing companies should be warned: unless they have sufficient resources to attain aggressive objectives and maintain a distinctive presence (such as by exclusive licensing of music talent or entertainment archives), they risk being swamped by the tides of disintermediation and consolidation.

E-Commerce

E-commerce can be simply defined as a trade of a product (goods or services) between a buyer and a seller that takes place over the Internet or other network. E-commerce business models are often divided into one of four main categories. (1) Trades that take place between one type of business and another: business-to-business, now universally shortened as B2B; (2) business-to-consumer, or B2C; (3) consumer-to-consumer, or C2C; and (4) consumer-to-business, or C2B.

In the context of this book, our aim is not to present a detailed treatise on e-commerce, but rather to compile a brief but useful overview of important e-commerce elements in the context of the online value chain.

E-Commerce: Competitive Factors

It goes without saying that, relative to traditional storefronts, electronic commerce adds value. Consumer benefits include lower prices, greater selection, increased comfort through home or office shopping, accessibility anytime from anywhere, reduced hours spent searching, quicker access to information and more detailed information and customization. Sometimes, however, due mainly to the seller's need to concede special pricing and provide for returns, the only participant that is clearly better off is the consumer.

B2C Benefits   This B2C model consists mostly of manufacturers or resellers that sell goods online. Operationally, these firms benefit from e-commerce in a number of significant ways, each of them a factor of reduction of bricks-and-mortar-type outlays. The benefits include reduced overheads, reduced sales staff, reduced cost of sales, reduced errors, reduced procurement costs through greater efficiency and selection, reduced inventories, reduced distribution costs (for digitizable goods) and reduced customer service expenses. One particular difference between e-commerce and off-the-street trade can be highly beneficial is the reduced or in some cases *no* working capital needs; in this model, the seller often gets paid before making a delivery and not infrequently even before making the product, or, especially in the case of resellers, before having to pay sources.

E-tailer Opportunities   The Internet also awards e-tailers with exceptional opportunities to enhance product lines, services and customer relations. Via *one-to-one marketing*, they can deliver sales pitches directly to shoppers whose preference profiles or previous buying habits have already been determined); or through *mass customization* they can aim sales campaigns offering personalized products or services at identifiable groups of likely buyers. They can identify, track and cater for their most valuable customers. They can increase cross-selling, reach new geographical markets and fine-tune the quality of customer service. Finally, they can effect considerable economies of scale and scope through wide product range and global reach.

E-tailer Challenges   Offsetting these benefits, e-tailers also face unusual challenges. Rivalry intensifies because the Internet facilitates channels of distribution that attract new competitors from other geographic regions and from other sectors, and triggers e-tailing entries into the market by brick-and-mortar retailers who might otherwise have remained in their store-fronts. Because electronic price comparisons are far easier than trekking through malls, the Internet can also be – and more often than not is – a demanding price-deflation mechanism. B2C customers also tend to use Internet searching technologies to find the best deals; thus they have low switching costs and harbour no particular loyalties. Fulfilment has proved to be a major challenge for non-digitizable goods, as has the costly construction and maintenance of a technical and logistical e-commerce infrastructure.

Channel Management   Storefront retailers who also are active e-tailers (some old-line chains merely put their catalogues online and hope for the best) face the vexing problem of managing different channels. Selling online conflicts with and – if the products are offered at lower-than-store prices, often cannibalizes – traditional, off-line distribution channels. Newcomers to e-tailing can move aggressively with no concern about such conflicts; however, they do face a major barrier to entry due to the comparatively huge collective investment required in branding, marketing and setting up sales systems, as well as in inevitably high customer-acquisition costs. As a result, e-tailers begin by spending more on each sale than they take in, especially after returns are accounted for. With venture capital having dried up in the early 2000s following e-tailing's dot-com fiasco, newcomers face all the more pressure to achieve profitability well before the cash runs out.

E-Commerce: Strategic Approaches

Strategies predominantly focus on (1) attracting and retaining customers, (2) adapting merchandizing methods that lead to increased order size and enhanced margins and (3) raising customers' switching costs to avoid perceived commoditization through price-only shopping. To attract and retain customers, firms are keeping records of customer's purchases and other available online histories. Based on these profiles, they send emails to targeted customers about related offerings; some dress up the offerings further with such value-adding lures as unbiased information and recommendations. Site management and order fulfilment must be outstanding, or customers will simply shop elsewhere. Amazon.com has generated high levels of repeat sales by creating network externalities and positive feedback[4] as its large customer base leads to more and more consumer book reviews and music recommendations, which in turn increase customer loyalty and attract yet more users. Three of Amazon's biggest threats are (1) retailers that offer their products cheaper in price comparison robots (by simply offering a discount over Amazon's price); (2) second the used market that in some of the core segments, such as books, offers a very comparable product sometimes at prices as low as 90 per cent discounted over MSR prices; (3) the threat from 'category killers' best-seller oriented local retailers that have less costs due to reduced inventory and faster delivery times due to their metropolitan locations.

E-Commerce: Industry Structure and Trends

Online Affiliates Both to avoid overt channel conflict and in response to competitive threats from rivals, some old-economy established firms have spun off separate online affiliates. Among them are bnbn.com (Barnes & Noble bookstores), and Toysrus.com/Amazon.com (Toys 'R' Us as a company-branded joint venture with Amazon.com). While the risk of channel conflict has been much publicized in the business press, many incumbent firms are finding, to the contrary, that they can actually leverage their physical presence by adding an online adjunct. The setup has been dubbed 'channel confluence'. Reaching the customer via different though complementary channels requires a progressive mode of retailing that some old-liners have not been prepared to embrace. In general, while such rethinking implies more convenience for the customer, it does present problems for the seller; one consumer-wooing stratagem, for example, allows a customer to return an online purchase to a physical store. The tradeoff is that it lowers the parent retailer's margins.

E-tail Consolidation Within certain e-tailing market segments, there may eventually be room for just one or two e-commerce firms. Web retailers whose business models are not distinctly different from better-established rivals will not survive. The drop-out process has already begun as the e-tailing industry consolidates. As much as 75 per cent of the B2C e-commerce audience time now passes through only five sites: Amazon, eBay, AOL, Yahoo! and Buy.com. In fact, in an emarketer survey eBay had over 30 per cent of the traffic of the top 100 US e-commerce sites. The speed, range, accessibility, and the low cost of handling information that network technology has enabled reduces transaction costs and thus stimulates new types of business activity and pricing mechanisms. Buyers and sellers are now able to participate in markets such as auctions and exchanges that with lesser technological advance would not have been feasible. (See Figure 8.5, p. 155.)

C2B Models In one C2B business model, some companies, such as Priceline, run reverse-auctions. In this case, transactions are turned around and consumers name their price for goods, such as an airline ticket; then sellers have to decide to accept the offer or not. However, this business model is something of a gimmick, inasmuch as the deal is defined by, and always works in favour of, the seller, such as an airline. Once the one-sidedness of the arrangement and the concomitant inelasticity of the market was demonstrated to investors, Priceline's stock fell from over $160 a share in August 1999 to under $5 about a year later.

Another C2B model is one that aggregates buyers in order to increase buying power and achieve the best possible prices – for example, Mercata.com. These new pricing models are expected to remain for products for which they work well, such as cars, travel and second-hand goods. In these markets, fixed prices are expected to increasingly give way to prices that reflect market conditions and address different kinds of customers with different variables.

Throughout the C2B realm, the Internet acts basically as a huge, free catalogue, excelling at attracting shoppers. This munificence is not always a bonanza, however, as the sellers are often poor at dealing with customers. Particularly misguided business models include the American dot.com food-provisioner WebVan which failed in 2000, essentially due to the erroneous supposition that if you offer a product, it will sell. Their distribution system involved constructing large warehouses before even one order had been taken. WebVan is worth mentioning since it is, as far as the authors have been able to research, the company that received the largest amount of capital in the seed phase. In Europe, an equally imprudent dot.com, UK fashion e-tailer Boo.com, was founded in 1999 and, having burned through some $126 million in venture capital, collapsed a year later (its brand and domain name were subsequently bought by a New York e-tailer). This was one of B2C's most prominent – and predictable – failures. The company did not provide for the fact that when a business sells a product out of a catalogue, eventually it suffers a certain percentage of returns. For e-tailers, that number can go as high as 50 per cent. To hope to stay in business, procedures must be established which open returned packages, check out the merchandise, repackage it and resell it. But Boo's two founders simply classified returns as lost inventory, inasmuch as the company lacked mechanisms to reprocess it.

7.2 Content Aggregators (Portals)

We can not follow on our analysis of the on-line value chain by looking at *aggregation*. 'Content aggregators' are companies that produce and manage sites that work by aggregating content of various other sites. The most well known are often grouped under the term 'portals'. These companies offer end-users an organized place to start their search and exploration of the Internet, linking them to Web sites according to their interests. Portals have consistently been the most visited sites on the Web (see Table 7.1).

Table 7.1 The top 25 properties, January 2003

Parent	Unique Audience	Reach %	Time per Person
1. AOL–Time Warner	78.569.257	64.26	6:03:46
2. Microsoft	74.226.727	60.71	1:43:38
3. Yahoo!	66.502.668	54.39	2:07:30
4. Google	29.354.436	24.01	0:17:21
5. eBay	28.430.707	23.25	1:36:29
6. United States Government	27.051.468	22.12	0:17:58
7. Amazon	25.930.103	21.21	0:15:25
8. Terra Lycos	23.626.076	19.32	0:19:17
9. About-Primedia	23.433.533	19.17	0:14:55
10. RealNetworks	21.224.869	17.36	0:18:27
11. USA Interactive	18.384.235	15.04	0:20:39
12. Viacom International	16.699.427	13.66	0:20:40
13. Walt Disney Internet Group	15.451.458	12.64	0:15:44
14. Sharman Networks	14.429.254	11.80	2:14:34
15. eUniverse	12.772.010	10.45	0:13:33
16. CNET Networks	12.711.255	10.40	0:11:01
17. InfoSpace Network	12.337.859	10.09	0:11:27
18. Landmark Communications	11.860.692	9.70	0:12:06
19. Apple Computer	11.717.307	9.58	0:06:30
20. Classmates	11.472.843	9.38	0:08:35
21. AT&T	10.415.095	8.52	0:33:05
22. The Gator Corporation	10.223.869	8.36	0:03:54
23. EarthLink	9.939.334	8.13	1:59:41
24. Ask Jeeves	9.295.451	7.60	0:09:36
25. WhenU	8.445.659	6.91	0:12:25

Source: Cyberatlas.

Early Portal Development

Early in the 1990s, the first horizontal, or 'pure' portals (general-interest portals covering a wide range of topics) were born as simple search engines or site directories. They offered Internet users an efficient way to filter the immense amount of information available on the Web. Since then, most horizontal portals (vs. scores of smaller vertical portals dedicated to particular subjects) have evolved into full-service hubs of electronic commerce, e-mail, chat communities and customized 'streaming' news. Often through a personalized start-page (today you can enter Yahoo!, for example, via a user-modified start page called 'MyYahoo!') they offer a place to start exploration of the Internet, linking to Web sites according to a user's interests. Since many Web-surfers start in directly at portal sites, these companies are in a powerful position, with vertical integration potential giving them considerable leverage over retailers and other firms that cannot afford *not* to be listed on their sites. Surprisingly, however, with the exception of a few highly popular portals such as Yahoo and those provided by

AOL–Time Warner these businesses generally have not been as profitable as was expected by many.

Portal Business Models

Aside from vertical portals, which are almost self-defining in that their focus is very narrow, there are two portal business models. One is the 'pure' portal such as America's Yahoo! and MSN, and Telefónica's Terra Lycos.[5] This type of portal is in a grey area, because the portal would not exist without the Internet. It isn't readily obvious whether they should be characterized as media companies or as Internet infrastructure providers. They possess many characteristics of media companies – high fixed costs, zero variable costs and highly active production departments – but for revenue they depend mainly on advertising. In any event, the 'pure' portal does not itself provide Internet access. It attracts an audience by packaging and promoting content (most of it non-proprietary), and generates revenues by selling both advertising (usually as intermittent rectangular banners) and 'anchor-tenant' fixed positions. Although this business model was once heralded as a paradigm of the new economy, Yahoo!, the biggest portal in the world, reported losses after having posted previous gains, and suffered a 90 per cent crash in the stock market (from over $200 US to under $20 in

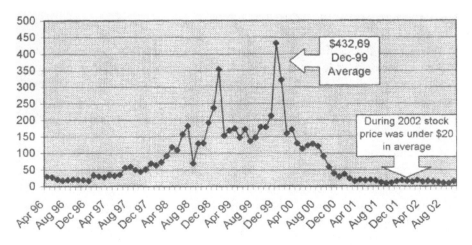

Figure 7.1   Yahoo!: life-time performance, April 1996–October 2002

Source: Yahoo!

two years; see Figure 7.1). The long-term strategy of the business model was thus placed in question.

Content Aggregators (Portals): Competitive Factors

On the value-creation side, portals have considerably reduced users' search costs. In addition, their customizable features, such as e-mail accounts, stock portfolio roundups and, selective news push created value still higher. Nonetheless, value appropriation is difficult, for several reasons. We will note four of them here.

Competition

First, rivalry among portals is intense with large, deep-pocketed firms competing in the general portal area and an increasing number of special-interest vertical enterprises, such as portals addressing only women (and at least one portal, About.com, that catalogues all vertical portals), entering the market. Once such companies have invested in the necessary infra-structure, they compete aggressively to build user bases to take advantage of the low variable costs of serving new customers. Mobility barriers among rivals are low, which is to say that for every idea or strategy that seems to be working, such as Yahoo!'s personalized start page, there are few ways to prevent the competition from copying them overnight.

Switching Costs

Second, despite portals' efforts to increase loyalty though customization, switching costs for visitors to portals are low and they move freely among them. Unlike IAPs, many of whom charge monthly fees to subscribers (especially in the USA), 'pure' portals such as Yahoo! have no contract to bind many of its users to their site. In addition, as users become more experienced with the Internet, they may migrate to more sophisticated or strongly focused vertical portals. Low switching costs also affect e-commerce revenue growth, since users who initially make a purchase by a portal may bypass the portal next time (cutting off its revenue) and go directly to that e-tailer's site (increasing its revenue). Thus a portal is easily disintermediated from the transaction process.

Technological Change

Third, technological change often requires new technological architecture within the portal. Because of low switching costs, it is critical to manage such changes well, as delays or interruption in service may encourage users to shift to competitive providers. Maintaining service is complicated by the fact that many portals depend on third parties for critical elements of their architecture. Firms spend considerable amounts of money and resources to provide a variety of communications services (e-mail, instant messaging, chat rooms, etc.). They provide these and similarly free inducements to users, but have not yet determined an effective means of generating revenues from them. Another consideration that can slow progress is that portals also must communicate with devices beyond the desktop, primarily to wireless devices. These devices require a different platform or version of the portals' service due to the smaller screen, lower resolution, limited functionality and undersized memory of non-PC devices.

Content

Fourth, one asset that cannot be imitated is unique and exclusive content. The ability to build and maintain productive partnerships will be critical to firms' future successes, as portals are becoming increasingly dependent upon third parties for much of their content, services and technological prowess. In efforts to differentiate themselves and increase customer loyalty, portals are investing millions (and in some cases billions) to obtain exclusive content. The battle for such content is in principle driving content prices up and availability down. Once they're set up and running, portals are *ipso facto* global businesses. However, unless they offer local content and language in other countries, they will struggle to transfer acceptance in their home markets to foreign markets.

Content Aggregators (Portals): Strategic Approaches

Ultimately, a portal's success will depend on its being able to generate maximum visitor traffic, thereby surmounting market transparency and lack of differentiation. This means satisfying a number of conditions. To maintain the free model among the unforgiving demands of portal competition, they must: (1) attract new visitors to their site, (2) keep visitors online at the site for as long as possible, (3) convince visitors to return (increase site 'stickiness'), (4) entice the type of visitors who will appeal to advertisers, (5) extract useful demographic and behavioural information from users and (6) encourage users to utilize as many of the portal's service and product offerings as possible.

The Customer's Experience

A major challenge to the growth, if not the very survival, of a portal is developing ways to enhance the customer's experience. If users cannot accomplish what they set out to do at a site, they go somewhere else. Therefore, portal content must be: (1) updated frequently, (2) of local interest, (3) easily and quickly accessible and (4) available on an increasing range of Internet-access devices. General (or 'pure') portals are maximizing the scope of their material so that the user doesn't have to leave or get redirected to another Web page. An important strategy for enhancing users' experience and achieving 'stickiness' is to concoct a 'community of visitors', a concept made possible by chat technology, for example. Indeed, many portals invite prominent guests to host the chat. In building a 'community', it is advised that portals compete via their systems, rather than their components.

Advertising

Advertising on the Internet still offers tremendous upside potential. Internet revenue from advertising currently accounts for less than 1 per cent of all advertising spending (online and offline combined) worldwide! Most Internet advertising is concentrated among a few of the top Web properties; however, simple arithmetic shows that smaller sites have gained their share of advertising – AOL and Yahoo! combined for a striking 55 per cent of the total in 1996. A growing source of such revenue comes from e-commerce (as opposed to advertising). The interactive advertising element of e-commerce – and thus the commissions that should be charged for directly producing the business – is made possible by Web tracking software that traces the source of the purchase to a given banner ad. While these commissions are still a relatively small percentage of a portal's total revenue (far more comes from space-billable advertisements themselves), commissions on banner-produced sales are increasing rapidly (see Exhibit 7.2).

Exhibit 7.2 Advertising Dramas

With the birth of dot.com companies, many pricing models for online ads were developed. The most popular of these models is CPM. *Webopedia* defined CPM as short for *cost per thousand* (the letter 'M' in the abbreviation is the Roman numeral for 1,000). CPM is used by Internet marketers to price ad banners. Sites that sell advertising will guarantee an advertiser a certain number of *impressions* (number of times an ad banner is downloaded and presumably seen by visitors), then set a rate based on that guarantee times the CPM rate. A Web site that has a CPM rate of $25 and guarantees advertisers 600,000 impressions will charge $15,000 ($25 x 600) for those advertisers' ad banner.

Since 2002, many companies focused on online market and ads research began to use a new measure: *Effective CPM*. Effective CPM relates to the approximate CPM value that you could receive when running ads that are performance-based. For example, if a banner pays on a per-click or per-lead basis, your effective CPM is a measure of how much you make on average for every 1,000 impressions delivered. It is not a guarantee of future performance, and does not represent a fixed rate, but can be used to weigh your options. If, for example, you serve 30,000 impressions to a CPA or CPC network, and earn $45 for doing so, your effective CPM is $1.50.

Having displayed very high levels of growth since 1996, both the online advertising market and Ads's price slowed down rapidly during 2001 (Figure 7.2). This slowdown is due to a combination of factors, including the contraction of the advertising economy as a whole, as well as problems intrinsic to the online advertising market.

The online ads' industry needs to address a number of problems, including the reluctance of advertisers to commit budgets to online media due to the perceived difficulty of measuring the success of online campaigns. The industry is addressing this issue by trying to

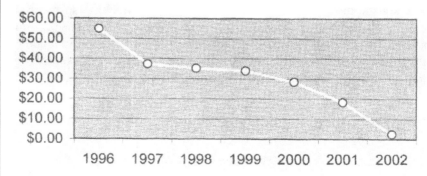

Figure 7.2   Online ads' price evolution (CPM measured), 1996–2002

Note: 2002 price is measured by the effective CPM.

increase the level of independent auditing of websites (at least one trade organisation introduced auditing by all its members during 2002). To supplement this, a growing number of publishers and advertisers are turning to traditional research methods, such as pre- and post-campaign awareness surveys, to monitor the success of their online activities.

As well as resistance among advertisers, there is also some consumer resistance to online advertising, due in part simply to the 'irritation factor'. In addition, there is unease among some consumers about privacy, especially since many ad servers track web users' movements to monitor the effectiveness of advertisements. European legislation is currently being proposed that would restrict the use of cookies (files used by websites to record data about users) – a move that is viewed with alarm by many in the industry.

The last reason explaining the slowdown of online ads' market is that consumers are tiring of banner advertisements – the standard format for online campaigns – and are now less likely than they were in 1999 either to notice or to click on them. However, this is also being addressed by the industry with the development of new standard advertisement formats.

Differentiating the Brand

Portal firms are spending considerable money and resources to establish, maintain and, often, to defend their brands (see Exhibit 7.3). At the same time, due to an increasing number of competitors, it has become more and more difficult and expensive for portals to obtain quality television, radio, magazine, Internet and other advertising space. So, as mentioned as a productive approach above, firms are expanding their services to find new ways of differentiating. Some are jumping into the corporate market by providing enterprise portal services, with pages similar, for example, to MyYahoo!'s, but focused on employees. Such portals are attempting to expand beyond the provision of content to the provision of online applications that offer Application Service Provider (ASP) solutions to needs such as e-mail, storefront presence and newsgroup management. Such tactics investments in these types of initiatives are proving to be progressively more important in establishing a differentiation advantage.

Strategic Alliances

Finally, it must be noted that because of the growing range of Internet devices online consistently, it's no longer merely an option to form strategic alliances with content providers. Now portals are being *forced* to partner with other companies to ensure a content presence across all devices. Sprint PCS, a major wireless operator in the USA, has signed multiyear contracts for content provision from both AOL and Yahoo! to ensure itself material for its mobile units, for example.

Exhibit 7.3 Google

Google is a play on the word 'googol', which was coined by Milton Sirotta, nephew of American mathematician Edward Kasner, to refer to the number represented by the numeral 1 followed by 100 zeros. A googol is a very large number. There isn't a googol of anything in the universe – not stars, not dust particles, not atoms. Google's use of the term reflects the company's mission to organize the immense, seemingly infinite amount of information available on the Web. Google's mission is to make the world's information universally accessible and useful.

The first step in that process was developing the best *search engine* for finding information online quickly and easily. The speed a surfer experiences can be attributed in part to the efficiency of Google's search algorithm and partly to the thousands of low-cost PCs they have networked together to create a superfast search engine. The heart of Google's software is PageRank™: Many search engines return results based on how often keywords appear in a website. Google is different. PageRank™ uses the vast link structure of the Web as an organizational tool. In essence, Google interprets a link from Page A to Page B as a 'vote' by Page A for Page B. Google assesses a page's importance by the votes it receives. Its also analyses the pages that cast the votes. Votes cast by pages that are themselves 'important' weigh more heavily and help to make other pages important. High-quality pages receive a higher PageRank™ and are ordered or ranked higher in the results. Google also combines PageRank™ with sophisticated text-matching techniques to find pages that are both important and relevant to the Web surfer's search.

To complement PageRank™ technology Google developed *Hypertext-Matching Analysis*. It analyses all the content on each web page and factors in fonts, subdivisions and the precise positions of all terms on the page. Google also analyses the content of neighbouring

Web pages. All of this data enables Google to return results that are more relevant to user queries. To support this innovative technology the company has the largest commercial Linux cluster with more than 10,000 servers linked all over the world.

Supported by their unique Web technology, Google's business model is based on two revenue streams: *search services* and *advertising programs*. Google's scalable search services, which include Google WebSearch™ and Google SiteSearch™, draw on Google's proprietary search technology and include a suite of fully automated options and capabilities. Today's major portals and corporate sites from all over the world, spanning all Internet platforms, have selected Google search services for their search technology requirements. Google WebSearch™ corporate customers include Yahoo! and its international properties; the three largest portals in Japan – Yahoo! Japan, Fujitsu NIFTY and NEC BIGLOBE – as well as NetEase (China), and Yam.com (Taiwan) in Asia; Vodafone Global Platform and Internet Services Group (UK), Retevision (Spain); and Sapo (Portugal) in Europe. Bellwether portal customers, including AOL/Netscape and the *Washington Post* in the USA, and Virgilio in Italy are long-time customers that have renewed their agreements with Google.

Google's advertising programs provide online advertisers with a choice of two text-based advertising programs: the Premium Sponsorship programme and the self-service AdWords program. Google Premium Sponsorship advertising delivers the company ad whenever someone searches on the keywords relevant to its product. It is a CPM-priced, full-service premium program that guarantees fixed placement for the ads in one of two enhanced text links appearing at the top of the Google results page. The CPM pricing is defined as the cost per 1,000 ads delivered. A company can calculate the cost of campaign by multiplying CPM × Number of impressions/1000 (number of impressions is defined as every time Google displays the company ad to a Google user). The minimum investment is $15.000 over a 3-month campaign (Figure 7.3).

Google AdWords™ enables a company to manage their own account, and with cost-per-click (CPC) pricing, is pays only when users click on their ad, regardless of how many times it's shown. The company controls its costs by setting a daily budget for what it is willing to spend each day. The service is available in 6 languages (Figure 7.4).

Figure 7.3 Google advertising program

Source: Google.

Today, Google is the principal search engine for the Internet in the world, with more than 200 million searches per day and more than 3 billon Web pages indexed. One of Google's latest moves has been to use its unique combination of database reach capabilities and massive predictive modelling skills to compete with Doubleclick (see Chapter 8).

Source: 'About us' on the Google Website: www.google.com.

Figure 7.4   Google AdWords™

Source: Google.

Content Aggregators (Portals): Industry Structure and Trends

Technology and industry *convergence* – the confluence of telephony, computers, the Internet, and television – is occuring at the portal stage as a result of forward integration (such as, among products and services, media companies or infomediaries) and backward integration (such as IAPs, telecoms, or browsers). Portals themselves are converging on other stages by offering Internet access (strengthening their IAP offerings) – Internet telephony, e-commerce and even hosting services. The industry with which portals are converging most rapidly is the online content stage (labelled generically in Figure 1.3 as Content Provider).

Owing to the importance of establishing a competitively meaningful installed base, firms moved as soon as feasible into international markets. The industry is consolidating as portals merge with one another – sometimes globally (for example, Spain's Terra acquiring America's Lycos in 2000 or the AOL–Time Warner merger in 2000, see Exhibit 1.1). As winning firms continue to extend their lead over the competition, it is increasingly likely that only a small number of firms will survive. The shakeout began in 2001 as first-tier portals (AOL, Yahoo!, MSN) captured from 70 per cent–80 per cent of site visits and ad revenues.[5] Because there was too little left over for them to run profitable businesses, second-tier portals (such as the now defunct At Home, which sold as high as $87 a share as late as 1999) rapidly faded from the scene. Nevertheless, despite these challenges and concomitant losses on the part of many participants, the function of the horizontal portal industry remains of strategic importance to its main players. This is because of its involvement with other, complementary, industries.

Notes

1. See J.E. Ricart, B. Subirana and J. Valor, *The Virtual Society*, Barcelona. 1997, for an extended taxonomy of products and services.
2. See L.C. Thurow, 'Needed: A New System of Intellectual Property Rights', *Harvard Business Review*, 75(5) 1997, pp. 94–103.
3. See also chapter 4 above and P. Evans and T. Wurster, *Blown to Bits. How the New Economics of Information Transform Strategy*, Boston, MA, Harvard Business School Press, 1999.
4. 'Network externalities', (also known as 'network effects' or 'positive consumption externalities)', as has been mentioned earlier in the book, exist when a user values a good or service more highly as the total number of users for that good or service increases. A common example is the fax machine; as more people own faxes, each user's machine becomes more valuable to that user; however,

eventually use of the good or service may become surfeited or, as in the case of facsimile technology, superannuated. In this case, network externalities no longer pertain as a market dynamic. 'Positive feedback' is an economic market force by which 'the strong get stronger and the weak get weaker' (C. Shapiro and H.R. Varian, *Information Rules – A Strategic Guide to the Network Economy*, Boston, MA, Harvard Business School Press). As W. Brian Arthur describes it, when a firm, product, or technology in a network industry gets ahead of the competition (whether by strategy or by chance), it tends to get further ahead, while the one which is behind tends to get further behind. Such markets are described as 'tippy' and often lead to a monopoly situation in which one firm takes over, or, in other words, locks-in most or all of the market. (Arthur, 'Positive Feedbacks in the Economy', *Scientific American*, 1990, pp. 92–9). In this context, Arthur cites the example of the 1980s struggle for IAP dominance among America Online, Prodigy and Compuserve which, as we have seen, was won by AOL despite Prodigy being the first to enter the market. Once AOL forged well ahead and created market momentum, its competitors had to drop out. (Arthur, 'Complexity and the Economy', *Science*, 2 April 1999, pp. 107–9). See also Chapter 2 on Client Software.

5. See also Sandra Sieber and Josep Valor, 'Market Bundling Strategies in the Horizontal Portal Industry', *International Journal of Electronic Commerce*, Summer 2003.
6. J. Black and O. Kharif, 'Second-Tier Portals: Going the Way of Go.com?', *Businessweek.com*, 31 January 2001.

CHAPTER 8

Extensions of the Value Chain and Transaction-Stream Electronic Markets

So far, we have concentrated on a linear information value-chain scenario mainly centred on the Internet and on media content. In this chapter we would like to give a broader perspective of what sources of information are available that may be subject to an information value-chain analysis. We also would like to show some of the limitations of the value chain by illustrating a particular type of electronic market, transaction stream markets, that require generalizing the concepts provided above to include situations in which multiple transactions occur simultaneously and in a non-linear form (i.e. in a network).

The chapter is divided into three sections. In section 8.1 we look at the different primary sources of information that can generate content to be distributed via corresponding information value chains. In section 8.2 we look at the role of information in business processes. Finally, in section 8.3, the longest of the chapter, we concentrate on transaction streams.

8.1 Information Everywhere

The Information Explosion

Since the World Wide Web was formed in the early 1990s, many new companies have been formed to take profitable advantage of a network's unparalleled ability to disseminate massive amounts of content to a multitude of interested parties. Some information is free and has no market value – government-collected data, for example, or intellectual works whose copyright has expired, or weather reports. But how much money does information that is not in the public domain, that belongs to one party and can be sold to another party such as to a subscriber to an online service – in short, 'content' – add up to in aggregate? No one really knows.

The world's total *yearly* production of print, film, optical and magnetic content would require roughly 1.5 billion gigabytes of storage. This is the equivalent of 250 megabytes per person for each man, woman, and child on earth. Printed documents of all kinds comprise only 0.003 per cent of the total. In 2000, the World Wide Web consisted of about 21 terabytes of static HTML pages (1 terabyte = 1 million megabytes, or the equivalent of the textual content of a million books like this one), and is growing at a rate of 100 per cent per year; 93 per cent of the information produced each year is stored in digital form. Of that amount, magnetic storage is by far the largest medium for storing information and is the most rapidly growing, with shipped hard drive capacity doubling every year. Magnetic storage is rapidly becoming the universal medium for information storage.[1] This is the conclusion of a study published in 2000 by faculty and students at the University of California at Berkeley in the USA. Yet, despite this almost unimaginable annual outpouring of ideas, facts, images, songs and mere idle chit-chat, not only have ICTs that distribute this deluge over electronic networks become both more versatile and less expensive in a remarkably short time, but the sources of information and the applications to which data are being put have grown more diverse.

Sources of Information

Illustrating this phenomenon is the multitude of directions that bodies of information have taken. We distinguish between *primary* and *secondary* sources of information. By primary sources of information we mean the actual device where the digital information is initially produced. For example, GPS devices located in trucks are the primary source of vehicle location. Secondary sources would be those that obtain the information from a primary source and structure it in a computer that is connected to a network so that others can use it. In Exhibit 8.1 we include a sampling of inventive 'data-grabber' devices and technologies that actively seek out information to put to use. Many of these applications are automated, and have brought significant cost reductions to the business processes that employ them. The potential in other industries for harnessing information through such automation, with similar opportunity for cost saving, is considerable.

Exhibit 8.1 Sources of Information

Here are some applications and components that can generate information. All these sources of information can be plugged into any content provider and unleash a specific value chain.

Geographic Position by Satellite (GPS)

In this application of information-gathering, receivers on earth use satellite signals to determine the exact physical location of a car, boat, person, or 'smart' bomb's target. In vehicles, a local road map may be superimposed on the data, enabling a voice or printed text to instruct the operator which street to take to get to a given address. Industry has adapted GPS in some esoteric ways – among them, to map crop rotation in farming and to calculate most efficient truck-delivery routes. Add availability messaging (mobile phone technology that can determine if the phone of a person you are calling is turned on), and a caller to a would-be recipient in a specially wired car will be able to tell not only who is driving that car but, though GPS, where and in what direction she is travelling.

Medical Alarms

Attached or implanted sensors can anticipate the need for responsive treatment or define the nature and extent of a patient's illness. Already in practice, an internally worn cardiac device transmits data which permits physicians to evaluate the patient over the Internet. For disorders such as high blood pressure, a data-transmitting mechanism could, in real time, warn of the onset of a stroke, which could then be treated within the critical first hour.

Radio Frequency Identification (RFID)

Responding to 'guns' that read the data stored in them, electronic tags (also called electronic labels, transponders, or code plates) identify the tag's user. Wireless data-collection technology can be applied, for example, at toll-collection stations on motorways in so-called 'fast lanes', which do not require either a gate, a human collector, or the need for a vehicle to slow down. They can also be used to track a product and its components throughout the supply chain.

Coded keys

Many hotels have adapted a system of encoded cards and doorway data readers for room keys. Not only does this protect the occupant through the capacity to be set to a different code for each rental time, but it helps smooth hotel management costs. Another benefit is that, using the same coded information, guests can be given access to special hotel promotions.

Remote Meter Readers

Gathering electric or gas usage figures for billing purposes or to monitor efficient distribution of energy can be done from inside a van traveling at 50kph. Each electric or gas meter in a neighbourhood is fitted with an interactive data-sending radio device that is 'woken up' by a signal from the van as it passes by. The device 'reads' the dials, and a computer in the van records the information. And the reader does not have to contend with angry dogs.

Data Mining, or Knowledge Discovery in Databases (KDD)

This concept is somewhat akin to that of megadata – data that affects other data (see Chapter 1). It has been defined as 'the non-trivial extraction of implicit, previously unknown, and potentially useful information from data.' The querying technique discovers novel patterns and trends within data that is not explicitly devoted to those patterns or trends.

Short Message Servicing (SMS)

SMS technologies receive and display e-mail-type text on a mobile telephone screen. By 2002 some 10 billion SMS communications per month were being sent to handsets through more than 240 mobile networks in about 100 countries. Dispensed cheaply, usually in real time, and sometimes through two-way interactivity, SMS material comes from mobile-content providers in sports, financial markets, news and similar sources.

The Supermarket of Tomorrow

An example of the benefits of applying information technology to business operations is suggested by two cost-saving techniques that will be adapted by a retail grocery of the future. When a good is purchased by a customer, a computer directive to replace it in inventory is automatically generated and sent to the warehouse. Depending on the data-mining response, the market will order more or less items per good sold, either to avoid overstocking the good or to avoid its being out of stock. Once the good is ordered and received, the computer-controlled warehouse stores it accordingly until the in-store computer 'sends' for it to put in the store.

The opposite case is when a good remains unsold. In this situation, after a predetermined period of time the system automatically transmits a signal to electronic price tags where the product is shelved. The new price proclaims the discounted level. To hone the procedure yet more finely, the store lends shoppers hunting for bargains a hand-held device. The PDA-like instrument reacts to overhead fluorescent lights, which not only illuminate the way but broadcast data onto a little LCD screen, such as product description, location and cost savings. (The shopper returns the receiver on leaving the store.) Despite the cost of installation and maintenance, this automatic guidance improves the grocery's profit and loss (P&L), since fewer unsold time-sensitive items must be written down to naught.

The Edible Camera

Under a technology called 'wireless endoscopy', gastro-intestinal-disorder diagnoses are made more quickly and accurately with the help of a capsule-sized video camera complete with its own light, batteries, transmitter, and antenna. The camera pokes around in the stomach and bowels of the patient, sending out valuable data to a recording device outside.

Embedded Chips

Described as a 'computer without a keyboard and monitor', a tiny microcontroller can be dedicated to real-time applications via coded 'firmware' that is etched into it. Embedded chips enable and/or monitor a wide range of ordinary events, like automobile-performance tests, console displays, washing machine cycles and shutter speeds in cameras. They can be so sophisticated as to help fly jet planes.

One kind of chip embedded in a so-called 'smart card', similar in shape and size to a typical automatic-cash card, stores and passively dispenses information regarding its bearer. A smart card not only carries standard particulars such as name and address, but can also be used as a universal banking card, thus obviating the need to carry several different cards. Most important, perhaps, in one application it can replicate on demand the bearer's unique facial, fingerprint, eyeball, and/or voice patterns, thus acting as a foolproof identification instrument. Adaptations such as this push out the limits of ubiquitous computing – and, of course, its corollary, the ubiquity of information.

But for certain applications, embedded-chip technology faces a daunting problem: they eat up power. Putting them to portable use away from plentiful power sources in such fanciful notions as wearable computers and non-stop mobile phone conversations, remain just that – fanciful except for yet single 1 kb, BFID taps. But in the near future, advances in lightweight but constant power generation will undoubtedly cure this handicap. Perhaps the juice will come from portable fuel cells, or from simply walking about, as electroactive polymers convert mechanical energy into electric power. Or even by a person wearing a coat tailored from treated silicon, which converts the sun's rays into a sufficient electric trickle.

8.2 Business Processes and Information

Interaction and Interactive Labour

From stock brokerage firms to bookstores, from farming to grocery delivery, *interaction* (defined by global consultant McKinsey & Company as 'the searching, coordinating, and monitoring that people and firms do when they exchange goods, services, or ideas') is radically altering the procedures of business. In 1997, an article in a McKinsey journal made a prediction based on rapidly changing technologies within the interactive channels. 'Interactions pervade all economies, particularly modern developed ones,' the authors said. 'They account for over a third of economic activity in the United States, for example. Interactions also exert a powerful influence on how industries are structured, how firms are organized, and how customers behave. And they are about to undergo dramatic change.'[2]

The piece goes on to note that the development of ICTs already is bringing about a significant increase in capacity and major reductions in costs, which in turn cause a spectacular increase in the potential for interaction. It is revealing to calculate the percentage of labour costs devoted to interactive activities, such as communications, data collection and communal problem-solving, as compared with the remaining labour costs assigned to activities that do not involve any interaction, such as physical labour, individual analysis and data processing. The McKinsey article traces the growing role of interaction. Tables 8.1 shows numbers for the USA, as of 1994.

The percentages in Table 8.1 do not break down the significant variations of interactive activity among the different activities and types of work carried out within an individual organization. The figure reaches 78 per cent for workers with interpersonal skills (such as doctors and teachers). The per centage is similar for workers in the retail sector, secretaries and

Table 8.1 Percentage of interactive activities, by sector, 1994

Area	Percentage interactive activities	Percentage non-interactive activities	GNP$ (billion)
Wholesalers and retailers	57	43	520
Personal services	54	46	1,36
Financial services and publishing	52	48	760
Transport, warehousing and public services	51	49	440
Manufacturing and construction	35	65	800
Agriculture and mining	35	65	120

supervisors. For senior executives the figure is 68 per cent, while for technical and scientific personnel it falls to 42 per cent. For people who work with data (such as analysts or administrators) the figure is 37 per cent. For nurses, waiters, hairdressers and other occupations of this type it is only 19 per cent and, finally, for machine operators, construction workers, drivers, etc., the figure falls to 15 per cent. Another measure that is remarkable is the steadiness of the shift toward interactive-type labour. These last two categories of worker above represented 83 per cent of US workers in 1900. This per centage has fallen steadily, from 70 per cent in 1930 to 51 per cent in 1960 and to only 38 per cent in 1994. Although these are rough estimates, we can safely surmise that the trends are ongoing.

Furthermore, these interactions rest on a substratum of information and are therefore highly affected by the technological changes taking place in ICTs, which continually deliver yet greater quantities and more useful information, as well as by the actual diffusion and penetration rates of these technologies into society in general, so that there is a wider familiarity with their application. Since the changes have continued unabated – and it's obvious that they are being publicly embraced by users of telephony that exploit the widening capabilities of wireless service, e-business protocols that are increasingly more sophisticated and precise, connectivity rates that have vastly increased in speed, and so on – we can conclude that the so-called 'information society' is a tangible reality, and represents a radical and substantial change in the basic elements of economic activity.

Interactive Business

Another major effect of the dispersal and penetration of the new technologies is the increase in business's skill at interacting. The economy is composed of an incalculable number of transactions, some occurring within business organisations and others occurring in the global market place. All these exchanges of goods, services and intellectual properties (IPs) are usually accompanied by interactions between individuals and/or organizations directed towards the search, coordination and supervision of activities involved in the ultimate transactions. The varieties of activities themselves are similarly incalculable – pricing, sales, inventory, warehousing, billing, customization, and so on. As we've seen, these kinds of information-based activities that support transactions and exchanges are becoming more and more predominant in modern economies.

Transaction Costs

Practically speaking, these interactions are the bases for identifying transaction costs. Transaction costs play a role in: (1) determining the limits of organizations by deciding which activities should be done internally and which should be outsourced, (2) selecting the best organizational structure from among the various alternatives, and (3) determining search costs in the markets and so determining the markets' efficiency.[3]

The McKinsey consultants calculated that the amount of time taken to locate a certificate of deposit with a given rate of interest by telephone is around 25 minutes; but if the search is conducted over the Internet, the time can be reduced to 1 minute, a savings of 96 per cent. Similarly significant reductions were captured in coordination-time reduction. Reordering an inventory item by mail, says McKinsey, takes some 3.7 minutes, by e-mail 1.6 minutes and via the technology of electronic data interchange (EDI), a mere 18 seconds – a final decrease of 92 per cent such low transaction costs not only are achievable, but the reductions achieved over a technologically sophisticated network are highly rewarding.

This fact alone will dictate major changes in the way commercial interaction is organized within the global economy. New markets have been and will continue to be developed in places where markets could not have existed before due to excessive transaction costs. (The existence of the entire electronic B2B industry, for a start, can be ascribed to this phenomenon.) At the same time, ICTs will bring about new methods and opportunities for strategic cooperation among organizations in extant markets, facilitating the integration of certain activities and the coordination of others that have not so far been integrated. Individual companies themselves will heed rigorously to realign their value chains. In short, virtually all commerce will be immersed in dramatic reshaping. The Boston Consulting Group (BCG)[4] has labelled this process 'Value Chain Deconstruction', signifying the amalgamated forces of destruction and reconstruction.

The Value-Chain Perspective

As we shall see in Chapter 9, we contend that one of the key applications of the value-chain framework is in *strategic framing*, i.e. in shaping the boundaries upon which subsequent strategic analysis can proceed. However, the value chain can also be used to understand what the value-added stages of internal applications are within a company. An example is the system of Mrs. Fields Cookies, whose IT information value chain is depicted in Figures 8.1 and 8.2. Mrs. Fields Cookies is a retail chain specialized in selling cookies. The information system is central to the business model of the company because most business processes relay on it. At the heart of the IT system is a database with all transactions, a workflow engine and a demand predictor that is used in the store to forecast staffing, procurement, production and other needs. The system also calculates suppliers' payments and payroll including incentives. It also helps in recruiting people by conducting standardized tests. The architecture that made Mrs. Fields famous is based on a central server that contains all the data and dispatches applications to the in store hardware once a day.

The point we want to make here is that the value chain can also be used to map different value-adding stages of internal systems. Companies that increase their level of IT outsourcing may find the framework useful in understanding the implications and possibilities. In the case of Mrs. Fields, the company opted for outsourcing much of their IT systems operation to a new company that it founded called ParkCity. This company is now providing services to many others in the retail industry, including Mrs. Fields competitors.

This example illustrates how a value-chain perspective can be applied to the analysis of information in business processes. Combined with process repository descriptions such as those of the *MIT Process Handbook*, it can also be useful in benchmarking the use of information across different processes.

8.3 Transaction-Stream Markets

The upshot of computer architectures that lead to networks like the Internet are extensive infrastructures that accommodate market agents scattered around the world. Collectively they enable the coordination and exchange of information. The Internet becomes an aggregator of such computer architectures, overcoming distances and lack of information, thus lowering entrance barriers and communication costs. This diminishing-cost trend is changing the nature of electronic markets by opening them to more and more fringe players. Given that costs of entry are considerably reduced, intermediaries are participating in markets they previously wouldn't have considered feasible, such as advertising, data security, digital notaries, digital currencies and retailing.

Unfortunately, at the same time such activity also fosters high security/trust costs. And trust, according to technology consultant Patricia Seybold, 'is the most important element for developing a virtual commu-

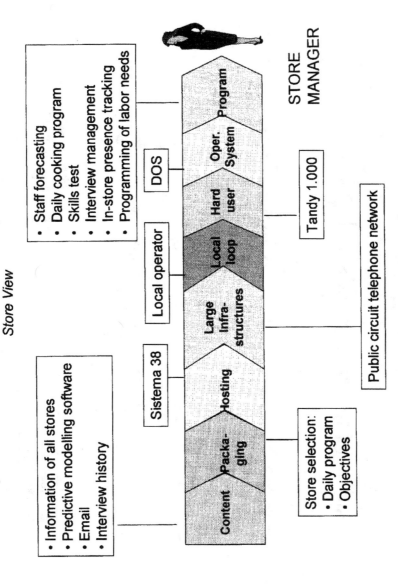

Steps of the Mrs. Fields Value Chain

Store View

- Information of all stores
- Predictive modelling software
- Email
- Interview history

- Staff forecasting
- Daily cooking program
- Skills test
- Interview management
- In-store presence tracking
- Programming of labor needs

Content

Packa-ging

Hosting

Large Infra-structures

Local loop

Hard user

Oper. System

Program

Sistema 38

Local operator

DOS

Tandy 1.000

Public circuit telephone network

Store selection:
- Daily program
- Objectives

STORE MANAGER

Figure 8.1 Mrs. Fields: value chain I

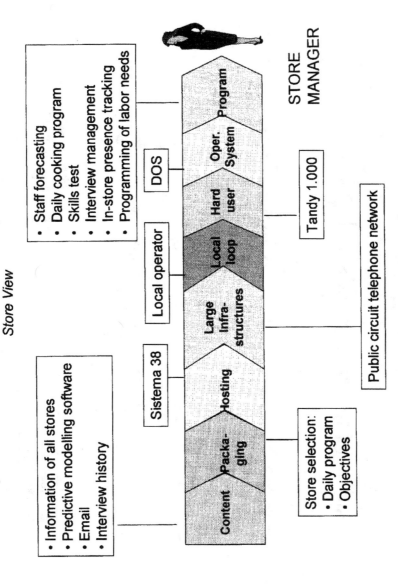

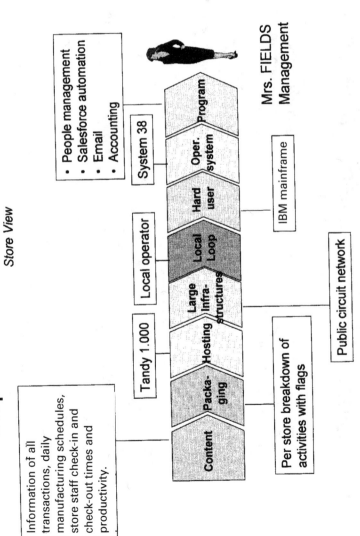

Steps of the Mrs. Fields Value Chain

Store View

- People management
- Salesforce automation
- Email
- Accounting

System 38

Mrs. FIELDS
Management

Information of all transactions, daily manufacturing schedules, store staff check-in and check-out times and productivity.

Tandy 1.000

Local operator

IBM mainframe

Per store breakdown of activities with flags

Public circuit network

Content · Packa-ging · Hosting · Large Infra-structures · Local Loop · Hard user · Oper. system · Program

Figure 8.2 Mrs. Fields: value chain II

nity'.[5] (A 'virtual community' can be defined in this context as a business setting in which a variety of persons outside that business make the decisions leading to a given transaction, with the business itself acting simply as the channel for communication.) In a similar vein, Electronic Data Service CEO Richard Brown noted in 2001 that 'the danger to the digital economy's longevity is not from the bursting of the dot.com bubble. Those effects are minuscule compared with those inflicted by breaches of trust'.[6] For computer systems that span multiple administrative boundaries, and especially on the Internet, such trust has proved difficult to establish. So far, global branding has been useful in providing assurance to customers. But otherwise, the lack of effective ways of managing the need for endorsement, security and insurance is a well recognized barrier to the full acceptance of transactions on the Internet.

Electronic Markets and the Search for Optimal Prices

Evolution of E-Markets

Economic observers have examined the differing roles of open supply-and-demand electronic markets and in-house managerial hierarchies as two discrete mechanisms for coordinating the transactions related to the flow of materials or services through adjacent steps in value-adding activities. In a forward-looking article in 1987, Malone, Yates and Benjamin[7] contended that the evolution of information technology, by reducing the costs of coordination (the management of dependencies among activities), was leading to a shift toward proportionately more use of markets compared to hierarchies to coordinate economic activity. They also stated that electronic markets are a more efficient form of coordination for certain classes of product transactions whose asset-specificity is low and whose products are easy to describe within an information-retrieval system.[8]

   In these authors' view, e-markets will evolve from single-source sales channels to biased markets (where the market maker acts as dealer, comprising one side of the market); from there to unbiased markets, in which the best available price is a product of best bid and best offer; and finally to personalized markets, in which the buyer uses customized criteria in making a choice. (For example, airline reservations systems allow the user to set preferences such as departure time, seat location, maximum number of stops, and so on; in such markets, each would-be buyer will probably stipulate different conditions.)

E-Market Components

In his 1991 analysis of influences on electronic markets, economist Yannis Bakos differentiated five e-market components – search costs (direct or indirect), increasing returns, switching costs, entry costs and maintenance costs.[9] He noted that a major impact of electronic market systems, in which information systems serve as intermediaries bringing buyer and seller together, is that (thanks to more comprehensive and faster browsing and/or the application of electronic robot comparison shopping) they typically reduce the search costs buyers must pay to obtain information about the prices and product offerings available in the market. This reduction in search costs plays a major role in determining the implications of these systems for market efficiency and competitive behaviour. The benefits realized do increase as more organizations join Web-channelled markets.

Bakos[10] anticipated that 'electronic marketplaces can impose significant switching costs on their participants'. But in fact, switching costs have been drastically reduced – or even rendered non-existent. The emergence of open standards for information transmission such as TCP/IP, HTML and Java – which prevented fragmentation of the Internet into incompatible models – and the concomitant overwhelming adaptation by all facets of market activities of the Internet as an open platform (on which, for example, the cost of a dedicated line is no longer an entry barrier) have been the engines of change. Java, for instance, can be used to create an interface that reduces switching costs by translating alternative interface options; thus, having to compile interface software is no longer necessarily a switching barrier. Ultimately, this phenomenon leads to an anti-intuitive outcome: as entry barriers diminish and numerous intermediaries do in fact enter, the complexity of e-markets increases. This makes it either very costly to compute an optimal price – or, at times, almost impossible as shown by Subirana.[11] We will expand on this further later in the chapter.

Observe that electronic markets can be considered within our information value chain framework because the transaction can be seen as a particular type of *content*, very much like the social interactive content of eBay as described on p. 18.

Transaction Processes

The Process Dynamics

IT systems have the potential for enhancing interactions among participants

by helping to leverage buying power and to streamline complex and inefficient processes leading to the execution of a transaction. A transaction is the
establishment of a contract between two or more agents to perform a given
economic action. Five processes comprise the completion of a digital transaction:[12] (1) player selection, (2) contract setting (3) contract signature, (4)
contract storage and (5) transaction action.

Figure 8.3 graphically represents this five-process dynamic. Let us take
a traveller booking an airline reservation over the Internet. The transaction
action involved is the transport of the passenger for a fee. In the process
toward achieving that end, a contract is established between the traveller
and the airline. In this or any digital transaction, the goal is to initiate,
arrange and complete these processes in the most efficient manner. To do
so, intermediaries can intercede between the two end-parties. In this case –
quite visible to the buyer – the most prominent would be a computer reservation service (CRS) such as Sabre, Amadeus, Travelocity, or Expedia.
Another intermediary – this one opaque – might be the security service that
safeguards the payment arrangements.

Figure 8.3   Transaction processes

Five Key Transaction Processes

Let us review each of the above transaction processes in turn:

(1) *Player selection* involves determining the economic agents that will be
    involved in the transaction. This is a critical component because it
    narrows the number of firms that will compete to get the contract to
    perform the action. CRSs refer only to airlines connected to their
    systems and, in a given transaction, display only those of that group
    that are relevant for the preset travel and itinerary conditions.
(2) In order to support current business practices as well as new ones on
    the Internet, electronic commerce needs the ability to enable negotia-
    tion between parties. *Contract condition setting* refers to the process by
    which the involved parties negotiate the details of the action that is to
    be performed. A CRS provides choices for given travel circumstances.

Information needed to determine the nature of a contract dictates the market mechanisms that will be involved in the other transaction processes. In other words, the process that details the clauses of the contract has implications for the way the market must operate. In the case of an airline contract to fly a passenger, the parameters and conditions are common and are simple to state and compare. Not so, however, in the equally common processes of purchasing a home. The complexity of that transaction, involving complicated contract and price negotiations, loan approvals, tax lending instruments, variable interest rates and formulas, and so on, has so far eluded electronic market models.

(3) *Contract signature* refers to the binding step in the process, in which the transaction players agree on a course of action that clarifies how the transaction activities will be performed. Often a market intermediary sets preconditions for at least one of the players to ensure the signature's validity. For example, a hotel reservation usually requires a credit card number to be held on file.

(4) *Contract storage*. In the conventional CRS, the contract – in this case, the ticket – is stored in the airline database, from whence it may be issued to the buyer either as a paper instrument or a virtual document.

(5) Finally we arrive at the *transaction action*, the process by which the activity referred to in the contract is executed. The airplane makes its transit, and the players on both sides of the negotiations are satisfied.

Transaction Streams

Transaction-Stream Dynamics

Transaction streams are e-markets in which more than one player is involved in at least one of the transaction processes (1)–(4) above. This means that various players are involved in the different transaction processes – a player may help in locating the parties to the transaction and another could set the contract conditions. In one case that has been widely duplicated (by search companies such as Altavista and Google, for example), a resourceful B2B revenue source surfaced on the Web in 1996, initiated by Amazon.com as one route to getting big fast. The concept was essentially fashioned after the venerable coupon-return system in print ads. Amazon dubbed it the 'Associates Program', recruiting for its members anyone who ran a Web site, from large companies to individuals. Amazon helped enlistees set up back-office operations, then shipped products and provided customer service for orders received through the special links that the members published within their sites. The sites' hyperlinks in effect led straight to Amazon's order desk, in essence populating the Internet with 'neighbourhood' Amazon outlets. Amazon rewarded them at commissions that ranged as high as 15 per cent. The instantly popular scheme allowed Amazon to access book and CD markets through over 300,000 web sites (as of 2000) at virtually no up-front cost. Just as significantly, the intermediaries suffered no entry costs at all! This transaction-stream example also demonstrates how the entry costs to some intermediaries can be close to zero.

There are thousands of bookstores on the Internet.[13] Their initial *modus operandi* appears fairly simple. An entrepreneur decides to make its database available on the Internet with the hope that customers will start surfing through his site and orders will land on his desk. This arrangement appears very similar to the sale of airline tickets through a traditional CRS, and is pretty much what Jeff Bezos (founder of Amazon) did when he left his job as a principal at D.E. Shaw in early 1994 to pursue Internet retailing. Music and books are some of the categories most suitable for Internet consumer retailing. There are more titles than any physical store can stock and a selection can be easily made by querying a title database. Books are also well-known commodities and easy to ship. In fact, in its first 30 days, Amazon shipped books to customers in all 50 US states and 45 countries around the world. The book market is also highly fragmented from the point of view of distribution, authoring and production. Some threat exists from large presence-based retailers such as Barnes & Noble and Borders because they may leverage their brand recognition with their warehouse and book logistics experience.

Through the Amazon Associates Program, Amazon provides the opportunity to link associate Internet sites to their own database. This means that the users of the associate site can purchase books pre-selected in the associate site. For example, in Netscape's developer area, there is a section with reviews about different books. The user can select a book and be redirected to the page that sells the book. Netscape has become an independent bookseller at practically no additional cost. The transaction profit is shared and Amazon gives a 5 per cent–15 per cent share to the 'Associates Program seller'. In 1998 it claimed to have over 150,000 sign-up associates and 2003 estimates point to more than 500,000.

Observe that the Associate Program changes two of the five processes described in the subsection 8.3.2. First, the player selection is performed by the associate partner, Netscape in our example. Then, in the condition setting, Netscape is performing a prescreening of the 2.5 million books of the Amazon catalogue. By sharing part of the revenue, Amazon is benefiting from a very inexpensive and enthusiastic sales force. This is very attractive for organizations such as Netscape that have access to a customer base interested in specific titles and permits access to the appropriate books when its customers want them. This can be considered a response by providers to reverse market trend.[14] To prevent the power shift to the customer, providers work together on a combined sales effort.

Entry costs are practically zero, given that Amazon provides all the back-office infrastructure; all that is needed to start a bookstore is a connection to Internet and a web-editor. What is the business of the associate? What is it selling? Observe that the key to the Associate is selling 'book-consulting'. Amazon provides a reward to such service in the form of the commission payback. The service is also very beneficial to Amazon given the high customer acquisition costs.

In fact, when Amazon buys advertising space in a search engine such as AltaVista, the combined effect is often that of converting AltaVista into a sort of up-front paid partner. Advertising banners can now be very targeted. For example, when a customer searches for the word 'Netscape', an Amazon banner can display the amount of Netscape-related books that are on its database. Furthermore, Amazon can create a banner with a search space so that the customer can choose to search either on the Altavista site or on the Amazon database. This case is interesting because the user is provided with alternative transaction paths while the firms involved are competing for her attention. The paper equivalent of directory search does not enable the dynamic player the selection and condition setting exhibited by successful Internet companies.

Altavista and other search engines such as Google also have an Associate Program to stimulate the usage of its database. Companies around the world are making specialized queries into the database. The logic to support such a program is that associate companies have better information about the user than Altavista or Google. This means that they can use such information to design better queries while retaining the customer attention. Observe that, here, Altavista and Google are involved only when the transaction action is performed. Since this program can be combined with the Amazon Associate Program, the number of players involved in each transaction multiplies. In fact, some of the queries delivered by Yahoo! are managed directly through an Associate Program by other players than AltaVista and Google at some point.

Transaction streams are electronic markets in which more than one organization controls the first four transaction processes (1)–(4). In the Amazon

Associate Program, the referring Internet site controls the player selection while Amazon controls the rest of the transaction processes. Underneath each process, many related actors take part and create more relations and transactions. Netscape is involved in the player selection and the contract condition setting. This transforms the traditional model described in Figure 1.3 to a series of *transactions streams*, in which more than one organization controls transaction processes (1)–(4).

As we have seen above, entry costs are practically zero in the Amazon case – in fact even lower than the revenues of the first transaction! The speed, range, accessibility and low cost of handling information that network technology has enabled reduces transaction costs and thus stimulates new types of business activity and pricing mechanisms. Buyers and sellers are now able to participate in markets such as auctions and exchanges that with lesser technological advance would not have been feasible.

Transaction-stream complexity helps explain the proliferation of intermediators that have been populating the Internet in an e-commerce setting that once was predicted to encourage the opposite trend – toward middleman and less *dis*intermediation (the Exhibit 8.2).

Exhibit 8.2 Transaction Streams: Basic Examples

In addition to advertising, there exist a variety of tactics which individual companies – especially B2C retailers – have adapted in an effort to broaden their scope, thus producing transaction streaming. Here are four of the most common:

Associate Inducements

As described in the Amazon example, this functionality allows companies to extend their reach by sharing audience and revenue with other Internet sites. For example, the spicy-food Internet retailer HotHotHot encourages third-party sites to provide links into its home page by offering 5 per cent of the revenue generated by customers brought in through these links. The HotHotHot associate is involved in player selection (by redirecting suitable customers) and in contract condition setting (by directing them to the relevant product page).

Special-Interest Links

Mail-order retailers have been particularly favoured by information technology. Before the arrival of electronic networks like the Internet, the player selection process used to be hit or miss. Print coupons in magazines were an uneconomical means of reaching special-interest customer types, and the data that directed direct mail was unreliable. Snail-mail marketing campaigns were considered successful if they garnered as much as a 2 per cent response. The Internet, however, allows transaction-activity players to trigger the player-selection process via links in specific Internet pages. Associate programs can be a way of sharing the revenue derived from such links; in one variation, the Associate program managing company also shared in some revenue by executing process (3) above, the contract signature.

Yellow Pages

The All-Internet Shopping Directory (a business founded in 1995) performs something like a Yellow Pages lookup by presenting a Web shopper with extensive listings of stores offering goods in such categories as flowers, computers, hobbies, services and B2B concerns. As with the old-fashioned *Yellow Pages*, this online directory provides various levels of listings, ranging from plain hyperlinks to more elaborate (and more expensive) focused listings and banner advertisements. The general category of such customer-grabbing companies is the pay-for-performance industry, where clients are charged a fee or commission for successfully performing a predetermined action such as a sale or customer lead. Some similar entities, such as the GTE Yellow Pages, host extensive product information and selling services, and get involved in the contract signature process.

Robots

So-called robots (they really aren't, of course) are programs that call on available search engines to scour the Internet, performing 'intelligent' tasks such as trying to find the cheapest price for a given product. (In the consumer arena they are known as 'shopbots' and take in banner advertising for revenue while being objective in their searches; one such shopbot is 'pricegrabber.com'.) Robots can be used to automate the functionality of all five transaction processes, but their basic involvement is in player selection and contract condition setting.

Transaction-Streams in Advertising

Advertising is another area where transaction streams are common. On the Internet, advertising is normally accomplished by placing banners (small advertisements which link to the advertiser's home page) on sites throughout the Net. When an advertiser wants to place banners on the Internet, it must select among a host of available media – in the tens of thousands in most cases. Such a selection is not as straightforward as selecting the airlines that are available to perform a given itinerary. Issues such as medium affinity, availability, etc. must be taken into account. Advertisers generally pay the publishers of the sites on which their banners appear. Individual advertisers, traditional ad agencies and cyber-ad agencies all aggressively purchase space for clients.

DoubleClick

An example of an active intermediary in online advertising is DoubleClick (founded in March 1996), which provides Internet advertising sales and management and manages an Internet advertising network. DoubleClick's objective is to bring its network sites, Internet users and Internet advertisers together. In order to do so, it has created a comprehensive database of Internet user and organization profiles, which are adapted for ad-campaign targeting purposes. DoubleClick proclaims two noble endeavours: 'improving the value of Internet advertising and keeping the Internet free for consumers'.[15] Its core business is selling advertisements inside commercial sites. They are not sold on the usual available-space basis, however, but rather with customized content aimed personally at the Web surfer who happens by. The strategy relies on software that tracks that surfer's Internet history and constructs a cookie-inspired profile of her preferences and interests; the ad then is able to deliver a personalized pitch to that individual. Additionally, DoubleClick categorizes every Internet page displaying DoubleClick ad banners in order to promote affinity targeting. Consequently, Doubleclick's rates tend to run comparatively high, since the response to a tailor-made appeal will presumably be greater than to a non-customized ad.

A DoubleClick Transaction

To illustrate the complexity of a Doubleclick enabled transaction let's take a trip with a request from an Internet surfer (let's call him e-Jo) as he connects to a site such as *The New York Times*. This appears to be a simple transaction, I ask you for a page, You send me a page with some ads. The point we want to make here is that even a simple transaction like this one can trigger a transaction stream where many players are involved. As we will see, when other players get involved, new transactions are generated that in turn trigger other transaction streams. This results in a cascading effect that vastly increases the number of relevant players involved in the simple transaction-stream example. This has profound implications for the structure and evolution of electronic commerce. But let us take our elevator trip to the Internet dungeons of the future (see Figure 8.4).[16]

Our trip will have 10 stops and the user may wish to follow it through Figure 8.4:

1 E-Jo sends a request to the *NYT* through the browser.
2 The *NYT* receives the request and prepares the content it wants to send to the user, possibly the home page.
3 The *NYT* also informs DoubleClick that there is a request by a user and that a banner needs to be sent.
4 DoubleClick contacts E-Jo's Internet Service Provider ISP and the ISP identifies the user. In fact, the ISP knows what other pages E-Jo has looked up, email and chat activity, etc. It knows all E-Jo's virtual behaviour. In fact it also knows his name, address, social security number and the bank through which E-Jo pays the ISP.
5 Armed with this information, DoubleClick contacts banks, hospitals, government databases, credit rating companies and other information providers in order to build a detailed user profile containing not only the information that the ISP has provided but also other 'non-virtual' information that is available throughout the economy.
6 DoubleClick, then organizes an instant auction and various companies respond. One of them is a predictive modelling company specialized in sports. It detects that E-Jo has an interest in basketball.
7 The predictive modelling company contacts Nike to decide what kind of product is best suited for E-Jo.
8 The company then looks at what areas within Amazon are best for E-Jo and the selected Nike product.
9 A deal is made between the predictive modelling company, the advertiser, the supplier and the information provider on the terms of the contract in case E-Jo does click in the banner. A digital notary registers such information.
10 DoubleClick is informed of the agreement and the banner is sent to E-Jo.

145

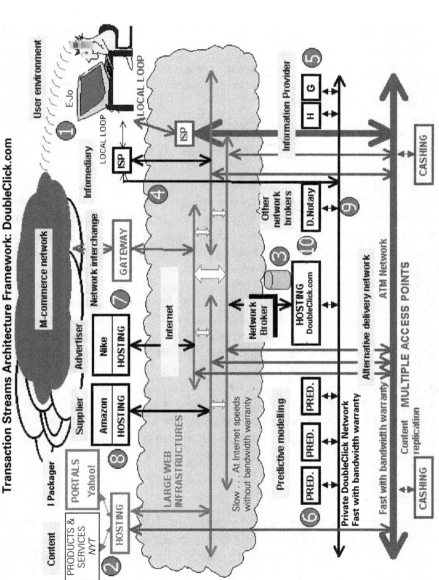

Figure 8.4 DoubleClick.com

Note that the example ignores the transfer of information back and forth between players. In each step more than four companies could be involved in this task – local loop, Web infrastructures, hosting and the connection to the server – and more than 40 players. We also ignore the possible connection via an m-commerce network, through a gateway to the Internet and a wireless local loop.

This should alert the reader to the increasing complexity of information activity. In the rest of this chapter we shall look at some of the implications of such increased complexity.

The Transaction-Stream Price-Search Problem (TSPSP)

Most organizations spend at least one-third of their overall budget on procuring goods and services. As one might expect, direct-line electronic markets in general demonstrate a trend to *disintermedation*, permitting customers to compare more providers in less time and thus lowering the search cost for the customer (and profit for the producer, due to the intense immediacy of the competition). However, modern markets on the Internet have brought about the opposite effect. Given their propensity for presenting intermediaries with essentially Free entrance as players in transaction procedures (as illustrated above in the Amazon Associate Program in Exhibit 8.2 and other transaction-stream examples), computer networks have become so complex that threading one's way through them to find the most advantageous price for a given good or service can exceed the network's ability to handle the search speedily and/or to produce usable results. In transaction-stream-based electronic markets, searching for the best price of intangible goods is what mathematicians refer to as an NP-complete problem (NP = 'non-deterministic polynomial time').

Computational Problems

The computational requirements for so-called 'NP-complete' problems are deceptively simple-sounding, but in practice can be unfeasibly demanding. Such problems require exponential time in order to find an optimal solution. One illustration of this phenomenon and of what exponential time means is a well-known classic poser that has earned its own acronym: TSP. The acronym stands for the Travelling Salesperson Problem, in which the challenge is to find the cheapest of a finite number of trips undertaken by the pretend salesman. A typical variation is as follows: given 100 scattered cities and the time of travel between each pair, find the shortest route for visiting all the cities, starting and ending at the same city and not repeating any. A foolproof approach is to draft a list of all possibilities, and choose the best solution. That, essentially, is what a computer search program attempts to do on a network. The trouble is that the number of possible routes is huge and the computation takes a lot of time – if, in fact, a computer can accomplish it at all.

In other words, solving a 'NP-complete' problem means that finding the best price will be increasingly complex as the size of the data to consider increases. The computational increase for NP-complete problems is such that linear increases in the input data result in exponential increases in the time needed to solve them. Even with abundant computational power, exponential demands soon defeat any existing computer. Finding an optimal route (in terms of distance or time) for the Travelling Sales person to visit a number of cities is NP-complete. From a practical point of view this means that one needs a simplified or semi-optimal algorithm to establish the route. It would not be difficult to write a program that finds the optimal solution by trying all possible alternatives. The issue is not whether the algorithm exists to compute the solution but what its computational requirements are. If one has 100 cities, visiting all of the alternatives requires us to visit the 100 factorial combinations (that is, 100*99*98*97*...*3*2*1) or 9,3*10exp157 – quite a large number. If we need 0,001 seconds to evaluate each alternative on a 500MHz machine, then the complete problem would be solved in 3*10 exp 147 years! What makes exponential problems hard to manage is that a little increase results in a much greater increase in the computational requirements. If instead of 100 we have 150 cities (an increase of just 50 per cent), the above example would yield computational requirements of 1.8*10 exp 252 years (an increase of 1,095 times!) Observe that there may be cases in which a trivial solution exists (or one without exponential requirements) – for example, if all the cities lie in a straight line. However, we do not know if there is an algorithm that can solve the most general Travelling Salesman Problem without exponential requirements.

There are many other problems that are of this nature and for which we do not know if there are non-exponential solutions. In fact, all of these problems belong to the NP-complete class of problems (or computational languages). One can prove that if there is a non-exponential solution for one of these problems, then there is a non-exponential solution for all of them. This is one of the most challenging open research problems in computer science, which is often quoted as the P=NP?[17] question.

How has such complexity come about? Within each of processes (1)–(4) described in Figure 8.3, many related actors take part, creating more relations – and transactions among them. Netscape is often involved in the player selection and the contract condition setting of a given transaction, thus transforming the simple Figure 8.3 model into a series of transaction streams in which more than one organization participates in at least transaction processes (1) and (2) (see Exhibit 8.3).

Exhibit 8.3 Infomediaries and Personal Information

At the Internet's beginning, observers understandably surmised that the networked interchanges of e-commerce would eradicate old-style middlemen; that, in short, it would bring about disintermediation. And, they further concluded, putting manufacturers and other basic sellers in direct contact with consumers would reward both parties with significant savings.

While some disintermediation did in fact occur, at the same time a powerful new variety of intermediary emerged. This type of intermediary, dubbed an 'infomediary',[18] performed a new function – looking after individuals' shopping and buying concerns. By gathering and providing objective company, product and consumer information, the infomediary greatly reduces asymmetries of information between buyer and seller. The resulting openness causes a shift in power from sellers to buyers by reducing customer switching costs (often to zero). It also tends to commoditize products and services, thus reducing the value of firms' brands. And this, in turn, increases price competition and reduces margins, forcing firms to enhance the efficiency of their businesses.

The most successful infomediaries tend to focus on solving a particular problem for a particular vertical market. By focusing on a distinct area such as advertising management or privacy control, infomediaries can consistently attract buyers and sellers whose primary interest lies in that particular area. If the infomediary function is sloppily conceived and lacks focus, many buyers and sellers will search for another site that does specialize on their specific topic or market. Intermediaries that are successful in building a loyal installed base can generate revenues by charging a percentage of sales to sellers for whom they facilitate transactions. They will also be successful in attracting advertisers interested in targeting a specific consumer market.

Personal Information

Before it was sold to Microsoft in 1998, a Web company called Firefly network (founded in 1996) claimed to have a database that contained profiles of more than 1 million Web users. The information had to do with each user's tastes and preferences regarding consumer products – music, electronics, movies, and so on. The idea at first was that through an online community, people could get in touch with other people who shared the same tastes and preferences. But the technology was such that it quickly went commercial, with Firefly licensing it to e-merchants, who could then tailor their server responses and aim sales messages directly at likely prospects. *Intelligent-agent technology* such as this can be used to lower search costs and can ensure anonymity for certain transactions; The intermediary can develop two separate trans-actions – one with the customer and one with the provider – such that the seller will not know the true name of its customer. Indeed, in this model of *reintermediation* (whereby the Internet is used to reorganize the buyers, sellers and other participants of a traditional supply chain), e-businesses adapting such technology would be involved in transac-tion processes (1)–(5).

Infomediaries

A relatively new Web phenomenon, *infomediaries* (from 'information intermediary') are neutral third parties who gather and organize content across a number of sources and functions as unbiased information aggregators for a target audience. In a typical application (there are a number of variations), the information is most apt to consist of a consumer's personal profile that typically is then sold to a merchant; the consumer protects her privacy by not revealing her identity directly to a merchant. In a typical C2B setup as performed by Autobytel, a Web site that specializes in helping to buy and sell automobiles, the consumer lists, say, her car-buying parameters with Autobytel, who then scours a network of some 5,000 dealers for sellers willing and able to fulfil the terms. Thus are consumers able to maximize the value of their personal information. And the infomediaries, by aggregating consumer information, become valuable intermediaries to merchants, who often pay commissions to infomediaries, so that they act much like brokers.

Predicting what the future of this new concept might bring, one Web infomediary, Lumeria, sees a reverse-image networked world in which 'in addition to merchants, I[dentity]-Commerce will include marketers and advertisers by allowing them to 'buy time' from individuals.' In it, an online shopper would register an interest in purchasing a certain product, such as a kitchen appliance. She would list descriptions of herself that would point the way to tailoring their sales pitches to the needs of this particular individual – and none other. The marketer then would compile a streaming-video commercial and a price offer which includes a customized rebate aimed specially at the individual shopper. In this perfect world, marketers would pay the individual not only for supplying her personal data, but for her promise to view the commercial.

Even with the aid of infomediaries and shopbots (such as mySimon or CNET), searching can be frustrating. One problem is that comprehensive search agents retrieve exhaustively long lists of hits and are not adept at screening them for relevance. Another problem is that the accompanying descriptions and categorizations tend to be inadequate. Lowest-price searches are all the more confounding (and ultimately more costly) for large firms, for whom even a slight difference in the price of a good or service can have a significant influence on competitive position.

That is why, in response to the ambiguous nature of much content, *metadata* (information about information) search techniques have become more widely applied. Such techniques (as applied, for example, by Yahoo!) are more efficient, but complications arise when prices are retrieved in standard metadata protocols. For example, true prices often are obscured by special offers, e-coupons, frequent-user discounts or brand-loyalty incentives. To refine the search process yet more effectively, new *lookup tools* have been described that help describe data more accurately, such as the Resource Description Framework (RDF), which integrates large varieties and quantities of information, and Extensible Markup Language (XML), an open standard designed to describe data, in a sense like html does, but aiming at the content instead of the container. XML was in 2000 expected to be the 'silver bullet' to cure B2B networking of its slow and costly turns.

Despite all this, increasingly complex networks frustrate a continuing downtrend in search costs. Linear increases in input data result in exponential increases in the time needed to 'visit all the cities'. The attraction to the Internet of an abundance of B2B, B2C and even C2B market players has brought about a proliferation of agents – that is to say, more and more intermediaries (who become involved in a given product or service fulfilment process. Add to this proliferation the increased number of market sources that can be electronically referenced, and the computational complexity of searching for the cheapest option becomes an NP-complete problem. Simply put, networks, especially the Internet, now foster an environment in which searching complexity increases, and the best price cannot be obtained in a timely, straightforward manner.

Dramatic Increase in Transaction Costs: Zero Entry Barriers in an NP-Complete World

At this point, two trends merge: the NP-complete problem creates a difficulty in browsing across the vastness of information data; and, the decrease in transaction costs reduce entry barriers to practically zero while extending the vastness of information within reach beyond the wildest dreams of the utmost compulsive information maniac.

Where should the balance lie? Somewhere beyond information intoxication, but not with search desperation and therefore abandon. The fact is that even if we have been blessed with the power to search more extensive and thoroughly, the means to browse overall human knowledge is still well beyond the single-clicked 'go!' button. Probably, in the future, search engines will continue to decrease the cost of transactions, refining one's will in order to deliver more accurate information, tilting the balance towards greater information access – But the struggle between the two trends will never be settled as long as we cannot find a way to solve NP-complete problems.

The Peer-to-Peer Model and the Evolution of Network Brokers

P2P Networking

Akin to the collaborative research mentioned above, peer-to-peer (P2P) is a manner of networking in which PCs communicate simultaneously among each other directly, rather than through a central server. This model (also known as person-to-person) is used partly because in large enterprises hundreds of desktop machines are often idle at any given time, leaving unused CPU and storage capacity sitting on desktops. The goal of P2P processing is to tap this potential by aggregating the resources across a network. This allows the PCs at the individual nodes to function collectively as one large supercomputer focused on a given processing job. (A 'job' in the P2P network mode, for instance, is the popular Internet-housed game 'Doom', which can simultaneously be contested among players scattered around the world.)

Napster

One manifestation of a P2P-like application came into cultural prominence in 1999 with a young company called Napster. Its software permitted people connected to the Internet to download audio files from each other's hard disks. Pursuing the founder's free-music-for-all philosophy, Napster's ingenious approach placed a huge amount of cross-computer content online, allowing users to swap hundreds of thousands of copyrighted recordings in the highly compressed (by a ratio of about 14:1) MP3 audio-file format. A typical popular-song track from a compact disc could be copied in about 1 second; in terms of storage, by 2002 a palm-sized, 6.5 oz portable MP3 player of such recordings was capable of storing about 1,000 songs – almost six days' worth of music! – on its internal 5 gigabyte hard drive.[19] Napster ceased practicing its free-tunes pursuit in 2001 following an injunction issued by a US court. But, in 2002, outdoing Napster and its MP3 file compression, a free music-sharing site called AudioGalaxy began trading entire CDs, offering not only all the disc's tunes compressed into one easily downloaded zip file, but the album's printed liner notes as well.

Pure P2P Networks

The ability of Napster's P2P network nodes collectively to generate several *terabytes* of storage and bandwidth at no user cost[20] was in its time a technological, if not legal, inspiration. But because it operates from its own centralized server, Napster does not qualify as a true P2P model. As Napster was intending to convert to a licensed, paid-subscription service, similar free-sharing software that did not employ a central server and therefore functioned as a genuine person-to-person system in that PCs communicated freely and directly among each other without central intervention, captured public fancy. Among such pure P2P Internet entities were Gnutella, Aimster, Bodetella and Navigator, each of whom eschewed relying on a single database that could be scrutinized and therefore controlled. Yet others, like Morpheus, BearShare and LimeWire, following Napster's socially generous initiative and cashing in on the low cost of storage and fast connectivity, moved into swapping files of all kinds and lengths, including pirated feature-length movie titles. And the Internet responded. Even while it was being sued by the US film industry early in 2002, Morpheus boasted over a million users around the globe online at any given time.

DoubleClick and eBay

Such *bona fide* P2P networks define one of the possible evolutions of the network broker model such as DoubleClick – and, perhaps, the future of networks themselves (see Figure 8.4). Some of the most recent moves of DoubleClick illustrate this. Its Associate Program allows advertising agencies and the like to create their own network broker model for their advertising customers. In turn, the software allows a subsequent level of Associate Program so that the advertising customers can have their own network broker. This means that, essentially, the DoubleClick network-centric model is evolving towards a P2P model. In this context, the DoubleClick business model is in part that of a technology solutions provider so that P2P can be enabled within the advertising context.

Other examples of network brokers displaying a transaction-stream-complexity model have been mentioned above: digital notaries, security and online currency exchanges. Add to them barter, gaming/gambling servers, Web-security providers and even collaborative research that in a supercomputer-like grid of networked PCs globally scattered participants such as scientists, engineers, programmers and contractors to work simultaneously in real time on the same project by using the Internet as their 'meeting place'[21] – and you still have a mere trace of intermediation potential. An example of such complexity is DoubleClick's enhanced network.

eBay is another network broker, perhaps the most successful, that has been able to generate a whole new market in the P2P space This is a widely cited example, not only because it attracts 30 per cent of worldwide page views of the top 100 e-commerce US sites (according to *e-marketer*) but because it illustrates the tremendous power of the network effect-system lock in-economies of scale (see Exhibit 8.4).

Exhibit 8.4 ebay

ebay's Model and the Monopoly Virtuous Cycle

In 1994, Pierre Omidyar was working as a software developer at a
mobile communications company in Silicon Valley. His fiancée, a serious collector of Pez paraphernalia, was frustrated over the difficulty in
locating other active traders in the candy's kitschy dispensers. Inspired
by his wife-to-be's inability to reach a fragmented market, Omidyar
devised a model for reliably bringing buyers and sellers together over
the Net, and founded eBay in 1995. One of the few endeavours of the
dot.com era to become profitable out of the gate, eBay evolved into the
Web's most popular and enduring auction site, and, since going public
in 1998, became a Wall Street darling. Omidyar's model is a lesson in
how to build and nurture a community through the creation of social
interactive content.

Customer-to-customer (C2C) sites on the Internet face a special
problem: the customers on either side of a transaction cannot be
presumed to be experienced in the methods of doing business. Both
have to trust each other – is the item genuinely what it's described to
be? Will it be shipped in a timely and safe manner? Is the buyer's
payment reliable? Such transactions must be somewhere grounded in
good faith, and indeed eBay's C2C concept was predicated on the
belief that people are basically good – a belief reinforced by the potentially punitive eye of the sponsoring site (not infrequently, eBay
banished participants who proved themselves disreputable).

Omidyar realized that the creation of a liberal sense of community
was crucial. eBay achieved this first by teaching customers to help
themselves, setting out the rules to which both buyer and seller were
expected to adhere, then by fostering confidence and trust not only in
the parties to the transaction (he devised a rating and referral system)
but, equally important, in the integrity and reachability of eBay itself).
Omidyar came to see 'community' (as many C2B sites have since
become known) in a broad, social sense – a life style, in effect. That is
why the site is realy an all-to-all commerce site (including B2B, B2C,
C2B and C2C). 'The eBay community is made up of a variety of
people: individual buyers and sellers, small businesses and even
Fortune 100 companies', says the description on the company's Internet
home page.[1] 'Large and small, these members come together on eBay

to do more than just buy or sell – they have fun, shop around, get to know one another and pitch in to help'. The establishment of trust and confidence in a company's e-commerce endeavour is crucial for negotiating the positive and negative externalities. Only when this is accomplished can the benefits of community reduce the transaction costs (of negotiating, monitoring and executing) in the delivery of convenience.

An endorsement of the underlying value proposition delivered by the eBay business model not only lies in the fact that the company was able to avoid stock market damage in the techno-slump of 2000–1, but that, unlike most entrants in the dot.com race, even during hard times it posted earnings growth of nearly 100 per cent annually in its first four years as a public company. By contrast, Europe's leading online auction house QXL.com continuously lost money. Figure 8.5 compares the price performance of the two companies common stocks during 12 economically troubled months ending in January 2002; QXL's nearly 90 per cent decline vs. eBay's approximately 20 per cent gain reflects investors' disillusionment with the former's less community-involved network-externalities-based model.

The largest C2C auction company, eBay, aggregates bidders and sellers around the world. It is one of only a handful of pure e-commerce firms that have been consistently profitable from inception, and is generally considered America's most successful Internet company. Bidding for some 3 million items per month, eBay's 10 million registered users average nearly 2 hours online every day. In addition, at around 7.5 per cent, eBay's commissions are considerably lower than the typical offline auctioneer's fee of over 25 per cent.[2]

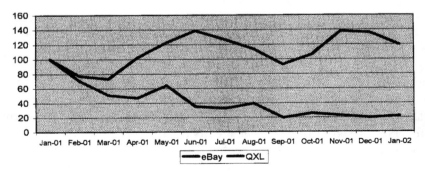

Figure 8.5   eBay: stock price evolution,
January 2001–January 2002
(January 2001 = 100)

According to eBay CEO Meg Whitman, the power of auction sites is their strong network effect: as the number of users grows, the sites become even more attractive and useful. In Internet commerce, it's a simple fact that people flock to where the action is. Ironically, this model of network externality is well illustrated by eBay's failure to dominate the Japanese market as it does in the USA. Indeed, in Japan eBay was such a distant second to market leader Yahoo!, which entered the Japanese market far earlier than eBay, that in 2002 eBay bowed out Japan entirely. Once Yahoo!'s network externality was in place, its dynamic effectively prevented eBay from gaining market share. (In a variation of the C2B model, the job-placement site Monster.com not only measurably ate into the help-wanted advertising lineage of printed newspapers, but became so well-attended by active browsers that its earnings began to outstrip eBay's starting in 2002, thus becoming one of the Internet's most successful dot.com companies.)

In addition to network effects, eBay is able to create significant customer lock-in by owning its history profile which is mingled with that of the other customers. Customer lock-in and network effects in turn give eBay the ability to leverage economies of scale. These three components create a virtuous cycle depicted in Figure 8.5 that propels it into monopoly situation – well, not in Japan where market results provide evidence that Yahoo! enjoys such a privilege. Note that this cycle is very similar to that described in various of the other stages covered, such as that of Microsoft in the OS segment of the information value chain.

Notes:
1. http://pages.ebay.com/community/aboutebay/community/index.html.
2. *The Economist*, 'E-Commerce', 26 February 2000, p. 5–34.

Microsoft

Microsoft is a company that is trying to create an infomediary that represents E-Jos on the Internet. Microsoft Passport is a service that collects user IDs (which in this context are similar to cookies and social security numbers), through a process that it calls activation; it also stores licence numbers of XP installations in a central database. The service could also be extended to .Net applications so that each software component on the Internet could have an identifier that was centrally located in Seattle. Furthermore, it could also be extended to the Product Internet (see below), so that each product on earth, including its components, had a centralized unique

ID in the Microsoft centralized Passport database. Let us now review the product Internet before we conclude the chapter.

The Product Internet

One of the key messages of this chapter is that the information value chain can be used in many other domains than in the Internet media example that drove us through the rest of the book. We have covered different primary sources of information (information everywhere), business processes and information and transaction streams. Here we would like to give a final example of where the combination of all of this may take us.

Enabled with Radio Frequency Identification Devices (RFIDs), each product can carry with it a unique ID.[22] With an RFID reader (essentially a wireless local loop), that information can be sent to a network broker so that transaction streams can operate in the product world outside of computers. Electronic agents executed in computers distributed throughout the world can represent the products and respond on their behalf (Figure 8.6).

Figure 8.6 illustrates an architecture of such a possible environment. It is strikingly similar to that used for DoubleClick (Figure 8.4). This is one of the points we would like to make – that the transaction stream architecture used in the DoubleClick example can be adapted to a wide range of other environments. Here the network broker is a company operating a virtual manufacturing environment, E-Jo is replaced by an E-Car or an E-Boat, or one of their components such as en E-Fabric. The local loop can be a WiFi/Bluetooth link from a fixed local loop or a fully wireless local loop linked with the Internet via a network interchange gateway. We still have advertisers, suppliers, packagers, content providers, predictive modelling companies and information providers. They are different companies with different focus than their DoubleClick counterparts, but the essence of the overall architecture is very similar.

Notes

1. School of Information Management and Systems, University of California at Berkeley, 'http://www.sims.berkeley.edu/research/projects/how-much-info/').
2. R. Butler *et al.*, 'A Revolution in Interaction', *McKinsey Quarterly*, 1, 1997, pp. 4–23.
3. See J.E. Ricart, B. Subirana and J. Valor, *La sociedad virtual*, Editorial Folio, Barcelona 1997.

158

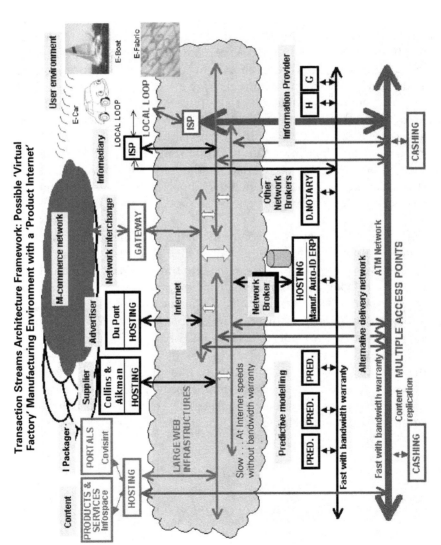

Figure 8.6 The virtual factory manufacturing environment with a product Internet

4. C.W. Stern, 'The Deconstruction of Value Chains', *Perspectives*, September 1998.
5. P. Seybold and R. Marshak, *Customers.com: How to Create a Profitable Business Strategy for the Internet and Beyond*, New York, Crown Business, 1998.
6. J. Schwartz, 'As Big PC Brother Watches, Users Encounter Frustration', *The New York Times*, September 5 2001.
7. T.W. Malone, J. Yates and R.I. Benjamin, 'Electronic Markets and Electronic Hierarchies', *Communications of the ACM*, June 1987.
8. In this context, 'transaction' can be defined as an identifiable operation carried out by or through an organization that transforms or converts an asset; an asset is said to be 'specific' if it makes a necessary contribution to the provision of a good or service and has a significantly lower value in alternative uses.
9. Yannis Bakos, 'Reducing Buyer Search Costs: Implications for Electronic Marketplaces,' *Management Science*, 43 (12), December 1997.
10. Yannis Bakos, 'Information Links and Electronic Marketplaces: Implications of Interorganizational Information Systems in Vertical Markets', *Journal of Management Information Systems*, 8(2), 1991.
11. B. Subirana, 'Zero Entry Barriers on a Computationally Complex World: Transaction Streams and the Complexity of the Digital Trade of Intangible Goods', *Journal of End-User Computing*, 1999.
12. Subirana (1999).
13. Case Study: 'Readers Inn: Virtual Distribution on the Internet and the Transformation of the Publishing Industry', Barcelona. IESE Business School, 1996. See B. Subirana, 'The Transformation of the Publishing Industry and Sustainable Business Models: Readers Inn, Cases "A" and "B" ', *Journal of Information Technology Cases and Applications, Quarterly Journal*, 1(4), 1999, pp. 51–74. See also 'Readers Inn: The Future of Product and Services Distribution and the Transformation of the Publishing Industry', *7th European Conference of Information Systems*, 3, pp. 964–83, Copenhagen, 23–25 June 1999.
14. Reverse markets is a trend by which the consumer power is increased by the use of technologies based on Internet, because he/she has access to a vast amount of information to compare and decide.
15. K. O'Connor, Doubleclick CEO, 2 March 2000; see also 'www.cdt.org/privacy/000302doubleclick.shtml'.
16. Many of the details of the example provided are not necessarily done by DoubleClick, but they illustrate some of the far-reaching potential that transaction streams can have in both the economy and society.
17. See also 'Zero Entry Barriers in an NP-Complete World: Transaction Streams and the Complexity of Electronic Markets', *IESE Research Paper*, 387, 1999.
18. J. Hagel, III and M. Singer, 'Unbundling the Corporation', *Harvard Business Review*, March–April 1999, pp. 133–41.
19. See 'How Stuff Works', http://www.howstuffworks.com.
20. C. Shirky, 'What is P2P ... and What Isn't?', November 24 2000; see http://www.openp2p.com/pub/a/p2p/2000/11/24/shirky1-whatisp2p.html.
21. See N. Hafner, 'Machine-Made Links Change the Way Minds Can Work Together', *The New York Times*, November 5 2001.
22. See www.autoidcenter.org.

CHAPTER 9

Conclusion: Strategic Framing and the Information Value Chain

9.1 The Relative Failure of Traditional Models

Strategic management tries to explain how value is created by the firm's activities and how this value reflects in the firm's performance. Strategic management as a discipline developed in the 1960s and 1970s. From there, in its short life, the strategic management field has developed many different frameworks to incorporate our additional understanding of firms and markets, as well as the changes in the competitive environment faced by firms.

Alternative Views of Strategy

Michael Porter developed a most influential framework 25 years ago. In his first book, *Competitive Strategy* (1980)[1] Porter taught us the importance of understanding industry structure to define competitive strategy. To do so, he proposed the use of the 5-forces model, i.e. the detailed understanding of (1) potential entrants, (2) substitutes, (3) rivalry, (4) suppliers and (5) customers as a way to grasp the structural characteristics of industries. Then firms compete by positioning themselves to exploit potential imperfections in industry structure. Porter complemented his model with his second book, *Competitive Advantage* (1985),[2] where he explained the differential characteristics a firm could use to make their positioning unique, difficult to imitate and, as such, a source of competitive advantage. The key competitive factor is to design the *value-chain activities* better to serve the target of clients.

As a reaction to this positioning framework, conceptually based on the industry–conduct–performance (ICP) paradigm developed in industrial economics, some researchers went back to the work of Edith Penrose on corporate growth,[3] where she emphasized the relevance of *corporate resources*. Several authors developed what today is known as the 'resource-based' view of the firm. From this perspective, the key strategic move is not to understand the industry and to choose a particular positioning to overcome competition; on the contrary, the key characteristic of a firm is its distinctive set of resources, capabilities and knowledge, different from its competitors, and difficult to emulate by them. The firm's strategy has to tell us which resources to develop, how to nurture them and how to deploy them in a particular product/market to sustain competitive advantage.

The 'Commitment' and 'Dynamic Capabilities' View of Strategy

From these two alternative views of strategy the field has evolved in different directions. Two of them are relevant to our approach to analyse the telecoms and information industries. First, is the need to develop a more dynamic view of strategy. Both industry and resource-based models are somewhat static. To overcome this, the so-called 'commitment' view of strategy was developed using concepts of game theory, where key competitive factors are the irreversible commitments that firms made throughout their history, which in turn are the basis of their stock of resources. In parallel, evolutionary economists developed a more incremental view, of 'dynamic capabilities', to highlight that the key asset of a firm is not its stock of resources and capabilities but its ability to generate a sophisticated response to the environmental shocks they face. In general terms, what the strategy field was doing was identifying different types of rents a firm could generate to create value: Marshalian rents, exploiting market power, by positioning; Ricardian rents, by exploiting differences in unique resources; Chamberlinian rents by making strategic commitments; or Schumpeterian rents by responding to environmental changes with new innovations.

The Value-Creation Model

In a second direction, and expanding the 5-forces model, some authors tried to highlight important aspects of the competitive environment that were novel or at least different from the prevailing context of strategy making. Authors like Brandenburger and Nalebuff[4] developed a value-creation model where the role of competitors, substitutes and potential entrants was augmented by the introduction of the positive force of 'complementors'.

'Complementors' are products and services that enhance the value of the company's own offering, increasing the value creates for its customers. Complementors, while perhaps not very important in some industrial products, are fundamental in information-based industries. This framework where firms can compete but also cooperate received the name of *co-opetition*. Hax and Wilde have developed what they call the 'Delta Model'. This model emphasizes *customer bonding*. The authors claim that firms have focused their strategies too much on creating the best product, forgetting stronger ways to develop a long-lasting customer relationship. They propose three basic positioning strategies: (1) best product, (2) customer solutions and (3) system lock-in.

Best Product

Best product is equivalent to the traditional competition based on cost leadership or product differentiation, where the key is to have superior performance in relation to competitors.

Customer Solutions

Customer solutions are where firms try to serve their current customers with an offer as ample as possible. The key to profitability is not overall market share of a particular product but the share in the needs of clients, sometimes called 'pocket share'. The basis for bonding is serving customers better than competitors, with a broad line, more personalization and (in most cases) understanding the needs of our client's clients.

System Lock-In

'System lock-in', creates the maximum bonding with clients because the key competitive factor is control over complements to activate network externalities or other types of positive feedback making defection by consumers almost impossible. As the environment gets increasingly networked, richer bonding opportunities are ever-more important. In system lock-in, profitability is a function of the share of complementors a firm masters. A huge share of complementors, like Windows currently has of software, makes it almost impossible for a client to switch from Windows to a different OS. If versions of the most popular software appeared written in JAVA to be run on any browser, or developed to work directly on LINUX, the dominance of Windows could disappear irrespective of the intrinsic qualities of Microsoft's operating system.

Strategies for the Future

In earlier chapters of this book we have reviewed some of the most successful strategies in information-based industries. Unfortunately, these models when used to study the competitive situation of the information-based industries (including telecoms, access providers, etc.) do not give us an understanding of possible strategies for success in the coming years. We have used them to analyse each sector, from content providers to software manufactures and telecoms operators, and have found that each one has structural characteristics that make them very unattractive and rarely profitable in the long run, except when a company dominates a standard. Open standards push entire sectors to commoditization, and we have seen extraordinary drops of profitability. Additionally, technology disruptions change the competitive landscape and what seemed sure bets a few years ago, like fibre optics in the home, are now questioned in the face of DSL, LDMS and Wi-Fi.

We can conclude that by-and-large these models do not help us understand the reasons for success or failure, and that they should at least contemplate the differential characteristics of the information product and the capabilities of the Internet. This is not to say that the traditional models cannot be applied to the telecoms and information-based industries, rather that they should be used with caution and explicitly take into account the five basic characteristics of these industries: (1) reduction of coordination costs, (2) reduction of search costs, (3) increasing returns to scale with almost no variable costs, (4) network externalities and (5) disappearance of the reach–richness tradeoff.

Factors (1)–(4) have been described extensively in earlier chapters. The disappearance of the reach–richness tradeoff refers to the fact that traditionally, companies have had to choose between systems that were either rich in information – i.e. had a high bandwidth and capacity for personalization or interactivity – or distributed a wide range of rather plain information. Today, this tradeoff has basically disappeared as technology permits massive distribution of fully customized information like the personal suggestions that Amazon.com provides to its customers. (See Chapter 7 and Exhibit 7.1 for some examples.)

This lack of comprehensive models has induced us to develop a new framework that incorporates the differential characteristics of the information economy and most of the traditional concepts strategy concepts. We have called this framework 'Strategic Framing' and it is based in our online value-chain model.

9.2 Strategic Framing in the Information Industry

The key element in strategic framing is the identification of where a firm is competing, what their business(es) is (are) and what activities are being incorporated in such business(es). We have used this framework to analyse and understand the competitive strategies of telecoms and other information-based industries. Strategic framing permits us to choose an adequate market bundling strategy looking for synergies across different steps of the value system, while changing the rules that regulate each market, as they work independently.

There are certain 'rules of the game' companies have to follow in order to compete in high-information industries. These markets feature specific characteristics that form the basis of the successful strategic approaches analysed below.

Compete or Collaborate?

One of the first decisions a business in such industries faces is whether to compete with differing standards knowing that one company may eventually take over the whole market share and leave everyone else stranded. Or – safer but less rewarding – to collaborate with fellow competitors and develop compatibility among systems and technologies for presentation to the market at a later stage. To highlight the particulars of this decision, we should apply 'co-opetition' strategy. This issue was underlined in the discussion on operating systems and browsers in Chapter 2.

Competing in high-information-content industries – whether new business concepts or translating old-liners such as stock brokerage, banking and grocery to the Internet – is what economist W. Brian Arthur has warned is 'gambling at the technology casino'.[6] In their search for a distinctive breakthrough, companies and their investors back what they hope will be superior technology. However, when positive feedback of some sort is present, speed in getting first to the market and planting your flag at the top of the hill – as, for example, bookseller Amazon.com did, can be more important than product or technological excellence. The rules for competing in these industries, therefore, have many variables.

For one thing, not all information-based industries are characterized at the same time or to the same degree by the key three conditions we have identified (high up-front costs, the potential for installed-base lock-in and the benefits of positive feedback). However, in whatever combination or degree, those are the key rubrics that define the rules for entering electronic-commerce markets from those that apply to whatever traditional industries remain.

Telecommunication companies are key competitors in this environment. Many of them evolving from the old integrated companies, called 'natural monopolies' because they evolved in a regulated and reasonably safe environment. However their competitive landscape has now changed as a consequence of the technological evolution. Deregulation and globalization have accelerated the pace of this change and many telecoms are now competing in many of the unattractive stages of the value chain like backbones and IAPs. But even worse, the rules of competition have changed: they now face a world of increasing returns to scale.

Strategic Framing

Strategic framing is critical in this environment. The main decision companies in the value chain face is to define *where they should compete*, and in the process how can they redefine the rules of competition to their advantage. The term 'Strategic Framing' has been chosen to highlight the fact that the real issue is how the problem is framed, we frame the problem by bundling stages better to compete against other different bundles.

Some examples are necessary to understand this concept. Recall, for instance, that ISP is a particularly difficult stage at which to compete because there are no barriers to entry, switching costs are very low, multiple competitors bring price towards marginal cost and recovering fixed cost is almost impossible. However a company such as AOL rapidly developed an installed base and created new products such as instant messaging that activated positive feedback to network externalities and essentially dominated Internet access in the late 1990s in the USA. To consolidate its position AOL used its power in the Internet boom and acquired Time Warner to get access to exclusive content and a cable distribution network. By bundling different stages AOL–Time Warner defined new rules of competition.

Information Industries

In this context of ever-increasing processing, storage and transmission capacities, a new industry is emerging on the ruins of the old telecoms unable to adapt to the new changes. The information industries, the content providers, the e-tailers, the informediaries, the many companies that are entering the online value chain to compete in it, with it and on it are The Barbarians of the information age. A key factor related to these companies is that they enter the fight with a less technological frame of mind. They bring content to generate businesses. They help in the reconfiguration of the value chain, transforming themselves as a key asset in any bundle. This is the basis of the demise of the telecoms industry and the rise of the information industries.

What are the limits of the information industries? Any business is today an information business. Dell produces, sells and delivers PCs, however the key element in its competitive advantage is the way it manages its virtual value chain, the ordering process, the dispatching and control of its operations and the software integration in its PCs. Dell competes differently, more efficiently because it understands the advantages of managing the virtual value chain that goes on the online value network.

Any business is forced to a greater or lesser degree to compete as an information business. Any business has to understand how to frame its information business in the context of the online value network. How can existing or newly founded companies identify and cope with the far-reaching changes that, despite the setback in pace (but not potential rewards) that the decimation of the 'dot.coms' brought about, are looming? We can identify three strategic responses – or rather, three facets of *one* strategic response:

- Rethink your business model and subcontract non-strategic and/or capital-intensive activities whenever you can.
- Narrow your focus toward being an intermediary – achieving a significant share of one or more of the stages of the fragmented value chain or of one of the intermediate transaction stream markets that ICTs have created.
- Frame your positioning by complementing your offer in one stage by having a sizable share in other stages so that you increase your stable sources of revenue, hopefully generating stickiness.

Let's consider each of these responses separately.

The Rethought Business Model

Fragmentation of the Value Chain   Yielding to the rush of ICTs, value chains have been fragmenting into finer and finer components. We have mentioned how once-humble Microsoft forged the first link in what rapidly became an incalculably extensive and constantly fragmenting value chain.

More and more companies are producing more and more specialized products, and (despite Microsoft's brush with America's antimonopoly laws) the end is not in sight. The same is true of Apple Computer, which started in a garage and directly and indirectly influences the products and services of tens of thousands of companies around the globe – many of them founded by Apple ex-employees. The scale of these chains is so far-reaching that one tends to forget the first links were forged not that many years ago, their rudimentary technologies almost instantly superseded by fresh innovation (the Apple II PC (1976) possessed a mere 4 kilobytes of random access memory (RAM), and no floppy disk or hard drive). The onrush of advance is breathtaking.

In that it opens fresh markets and diverse paths to extant ones, fragmentation of the value chain affords opportunities for companies to identify and adapt new ways of performing old business, putting old-school routines on notice. Amazon.com, in organizing an innovative online approach to bookselling, caused storefront booksellers to reconsider their classical style of retailing. Barnes & Noble set itself apart from e-commerce by introducing in-store coffeehouses that invited customers to linger and physically examine the merchandise. But that wasn't enough. The chain also started a separate e-business, (barnesandnoble.com), to compete – so far unsuccessfully – with Amazon.com.

Uncoupling Stock and Range   Nonetheless, ICTs enable bricks-and-mortar merchants to modify their operations successfully. There is no longer a need to strike a balance between holding too large a stock of goods, with the capital cost and risk of obsolescence this entails, and having a sufficient range of items on display for customers to choose from and buy. This used to mean finding just the right level of stock for the store and to operate convenient warehouses from which to fill orders and store excess merchandise. Now it is possible to uncouple these two activities. Maintaining the physical stock in the warehouse can become more efficient by finding an outside supplier who can store it and deliver it at the speed of delivery required, while at the same time the desired retailing state of having a wide range of goods on display can be simulated via a computerized catalogue made available to customers to search and order from. But such uncoupling makes it necessary to optimize the cost of each activity. This is true even when a business remains integrated and operates in several steps of the value chain and seems to be running smoothly. If its competitors are optimized on the basis of a superior business model, an integrated company may quickly find itself in danger of extinction.

'De-Averaging' and 'Deconstruction'   One change any integrated business challenged by competition can institute is the elimination of certain tradeoffs that for decades were implicit in the traditional value chain. The Boston Consulting Group (BCG) calls this process 'de-averaging'.[7] Companies that combine activities and their costs to achieve a competitive advantage are finding that fragmentation of the value chain is causing this advantage to disappear. Now such companies need carefully to attune their cost to each activity, instead of relying on tradeoffs to smooth the curve. They must find ways to outsource some *core* activities (such as ware-housing), not just marginal and/or support activities. If there is an operation that your company doesn't have the capability to process efficiently, your value proposition becomes better when you outsource it. The BCG calls this important notion 'deconstruction' – a reduction of destroying + reconstruction.

Of course, you lose some control, but that's a deal companies are willing to make. Before, control was everything: the business world was integrated into market sectors within which companies could comfortably operate their subdivisions. But now integrated businesses have to compete against eversmaller pieces of the market. Before it's too late, each division of the business must come to grips with the defragmention of the value chain. Fortunately, the number of strategic options has increased markedly. While it is too simplistic to go from one extreme to another (if only because there will be high coordination costs), perhaps the best concept is to concentrate on one or two things, and jettison others. In responding to new competitors you must rebuild your model so that your product or service stands out as an attractive value proposition within the defragmenting market place.

Uncoupling the Value Chain   Uncoupling the activities of the value chain is fundamental, since each uncoupled stage may proffer a separate business opportunity with its own sources of competitive advantage. Let's take a fictional metal-distribution and components milling company that has carved out a living by operating as a buffer between inefficient manufacturers and their wholesale customers. Its value-chained activities include sales promotions, order reception, metals purchasing and brokering, materials logistics and contract milling. All that sounds well and good, but in fact such a business should be organized more effectively. For one thing, the logistics activities should be outsourced, utilizing a single computerized central warehouse which would be far more efficient than the disjunct locations the company now physically maintains. And sales promotions should be uncoupled from marketing campaigns by installing a computerized telephone system and digitized order-taking facility. For this company, the two uncoupled functions that have the greatest prospect of becoming separate businesses with competitive advantage are metals brokering and parts fabrication; the more troublesome metals-purchasing component and its attendant customers might be dropped altogether. The strategy is most effective when a business uncouples physical activities (parts fabrication in the example) from information activities (marketing and brokering), as appears likely to happen with many businesses, since information activities will be able to take advantage of network economies and take on a high-per cent-of-the-market monopolistic cast.

If opportunities arise to design a more efficient business model utilizing ITC technologies, which they will as value chains become fragmented, businesses which fail to react will suffer serious consequences. A recent case in the USA is that of 53-year-old Polaroid, once an exciting part of the photography and film industry and a member of Wall Street's safe-invest-ments' 'Nifty Fifty'; failing to respond as photographic technology branched out into digitized imagery adaptable to the Internet, Polaroid went bankrupt in 2001.

Responding to the Challenge   In sum, responding to the challenge of remodelling a business require three things:

- Devising various scenarios for *de-averaging your operations*. Appraise your competitive advantage in each by studying the business models of your direct, indirect and potential competitors, in order to understand which of your own activities has the greatest promise of succeeding by being uncoupled.
- Understanding where and why your current business model has become *vulnerable* in the information-intensive economy (where marginal cost is more important because the relative effect of fixed cost is less important), what options to strengthen those points are available and how fast you can make the changes.
- Identifying the *internal barriers* that may impede this change (for example, being too enamoured of your management structure) and what you can do to overcome them.

Although it requires internal reorganization, outsourcing can resolve remodelling roadblocks.[8] A stratagem in the breaking up of the value chain is therefore to avoid activities that could be considered non-strategic, or that require major capital investment; if the company is already involved in such activities, plan to diminish their role or drop them altogether. The goal is to orchestrate a network, as, for example, Dell, Nike and Benetton, among other enlightened corporations do. These are cases where the network is focused on providing an excellent product or service to a broad customer group by controlling some (but not all) of the basic stages of the value chain. In Dell's case, that control aims at providing reliable, budget-priced, customizable-on-order, quickly delivered products that are sold over the Internet, by telephone, and through catalogues; embracing a fragmented part of the market, Dell also delves into financing, offering a variety of payment options to both businesses and home users. Basic stages of the PC value chain that Dell does not embrace are operating its own retail stores or issuing long-term warranties, both of which its less successful competitor Gateway has tried. Dell's value-chain management has paid off, in part due to its ability to survive low-margin price wars, in which its rival Hewlett Packard essentially abandoned the home- and business-systems markets entirely and moved more decisively into the value-chain fragments of PC peripherals such as printers, scanners and inks.

Compiling a network by controlling basic, non-capital-intensive, parts of the value chain is difficult to implement, and even more difficult to sustain, given the rapid changes that inevitably take place owing to information technologies. Today, value chains are intricate networks of interactions, many of them electronic, within which nothing can be taken for granted. An offshore supplier developed by a company specifically to serve that company's needs may seek other outlets for its goods if there are compelling alternatives. Or the supplier may run into financial difficulties of its own. Or the overseas supplier's quality control may not be satisfactory. Once having lost competitive advantage, it's expensive to regain it, as Nike found after its earnings per share were cut in half owing to unanticipated supply problems.

Having painstakingly assembled a network, to protect your position you must ensure that your company has sufficient resources which are difficult to imitate by competitors, and/or possess valuable assets that are not easy to steal (such as brand equity), and/or can develop new capabilities more quickly than its rivals, and/or be able to weather margin squeezes in the heat of price competition. An aggressive, price-conscious, and relentless merchandiser, Wal-Mart is a good example of such defences.

Being an Intermediary in the Fragmented Value Chain

If the preceding strategies strike readers as risky business, they need not worry. The fragmentation of the linkages within the traditional value chain infact enables a conservative strategy. Instead of attempting to orchestrate a network or radically alter your business model, you can now focus on achieving excellence as one part of a fragmenting value chain. Or, similarly, establish a niche market by anticipating and subsequently controlling one of the emerging intermediary markets in the way eBay and DoubleClick have done. To do this, you need to calculate accurately the soundness of five key propositions:

- The activity must be substantial enough to be *profitable in isolation*.
- The stage of emergence must have *value-creation potential*. The most important measures of potential markets are those that are sufficiently sensitive to the scale of supply and/or demand for the winner to be able to corner the lion's share of the market. The winner in online book retailing was clearly Amazon.com, which maintained king-of-the-hill invulnerability against all comers, even through the dark days of the decimation of the dot.coms.
- You must achieve the ability to *create value before your competitors do*. Again, see Amazon.com.
- You must be able to *defend your competitive position* once you have created it. More often than not, this requires having enough cash available – not usually a simple matter.
- The activity must offer potential for *extending the business to other related activities*, even though they may be in different sectors. Any number of companies have understood this – but, unfortunately, not all with the same clarity of vision.

Framing your Positioning in a Set of Stages of the Value Chain

The so far seemingly most successful strategy in the information industry has been to secure the best of the different stages of the value chain. Companies like AOL–Time Warner and Telefónica, not to mention Microsoft, have decided to compete in a number of intrinsically non-attractive stages and in early 2003 are doing better both in profits and market valuation than their counterparts. By combining IAP activities with horizontal portal, proprietary content, and in some cases the browser, companies can provide the 'stickines's of the content, the differential speed of the IAP for that content and the indexing capabilities of the portal and browser. If successful, this strategy will generate recurrent visits enabling targeted advertising in addition to the revenues of the IAP, while greatly increasing switching costs and making competition much more difficult for the companies operating in a single step of the value system.

The (at least temporary) successful strategy of these strategies is depicted in Figure 9.1, which presents the different positioning of the companies we have discussed in this book.

Vertical Integration   The portal property with the most visitors is the group AOL-Time Warner, the company that competes throughout the entire value system, from browser to proprietary content (including its own IAP and infrastructure from Time Cable, as well as its own portal). We can say that AOL follows the traditional rules of vertical integrations: eliminate intermediaries, exploit economies of scope, and leverage the brand in different industries. Also, as Saloner and Spence[9] have stated, the company leverage new economy factors like increasing network externalities through synergies across different steps of the value chain.

Integration also allows doing bundled pricing. Users need an IAP, and they will shop around for the best price: quality ratio. If a particular IAP provides more than just access, including some content that the user is willing to pay for, such an IAP is likely to get the user's access business.

Market-Focused Strategies   In contrast, those companies that follow market-focused strategies have run into trouble. In particular, offering technology alone is not a sustainable business proposition.[10] Consider Altavista and Excite@home: these two companies competed with technology in two different industries, search engines and broadband IAP. Their inability to be permanently the superior technology and the lack of 'stickiness' of their proposal made them disappear in the face of better technology (Google for Altavista) or cheaper cost/bandwidth (ADSL for Excite@home).

Value-Capturing Strategy   One could consider that the most value-capturing strategy would be one that included chat and communication services within the IAP–portal bundle, making it impossible for non-clients to access the services. This risky strategy has the most powerful network externalities, and if a company could establish itself as the largest IAP chat-messaging provider, it would undoubtedly dislodge all competition and reap enormous profits. Just consider Microsoft's strategy with its Messenger, Hotmail and Passport. If these services become prevalent, one could imagine a move that made them available only to users that switched to MSN.com as their IAP, and therefore capturing that part of value from the system as well.

9.3 Preparing for the Future

As they say, 'the future starts today'. Technology has changed the rules of competition in the twenty-first century. We are in a great need of understanding its implication. Business and society are evolving trying to use,

173

Different Strategies of Portals and IAPs

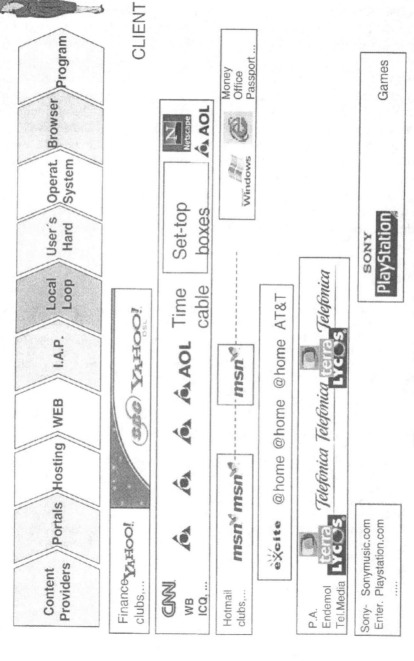

Figure 9.1 Value chain with positioning

understand and leverage the technologies of today and those emerging ones that will dominate tomorrow. Technology changes the structure of industries, fragments the value chain and creates new ways to interact with costumers and organize economic activities. In this complex, turbulent environment telecom companies transform into information businesses, but even worse, all businesses are in some degree information businesses. So we are all competing and consuming in the information industry. And in this environment old frameworks need a new frame. This book provides such a new frame. Thanks to the online value chain and the profound understanding of the rules of competition at each stage, all business can better frame their strategies acting at two levels: (1) understanding the evolution of competition in the online value chain, (2) deciding how one can use this understanding to reinvent ones own business.

Our model is just a tool. A tool to re-invent your business. A tool to create a better future. A tool to frame your strategy so that you ask yourself the right questions. The first step to reach the right answers.

Notes

1. Michael E. Porter, *Competitive Strategy: Techniques Analyzing Industries and Competitors*, New York, Free Press, 1980.
2. Michael E. Porter *Competitive Advantage: Creating and Sustaing Superior Performance*, New York, Free Press, 1985.
3. E.T. Penrose, *The Theory of the Growth of the Firms*, New York, John Wiley, 1958.
4. A.M. Brandenburger and B.J. Nalebuff, *Co-opetition*, Boston, MA, Harvard Business School Press, 1996.
5. A.C. Hax and D.L. Wilde, *The Delta Project*, London, Palgrave, 2001.
6. W.B. Arthur, 'Increasing Returns and the New World of Business', *Harvard Business Review*, July–August 1996, pp. 100–9.
7. Evans, P.B., 'How Deconstruction Drives De-averaging', *Perspectives*, October 1998.
8. See also D.C. Edelman, 'Patterns of Deconstruction: The Orchestrator', *Perspectives*, November 1998.
9. G. Saloner and A.M. Spence, *Creating and Capturing Value: Perspectives and Cases on Electronic Commerce*, New York, John Wiley, 2002.
10. M. Donegan, 'Contemplating Portal Strategies', *Telecommunications*, 34(2), 2000, pp. 48–52.

W.B. Arthur, 'Increasing Returns and the New World of Business', *Harvard Business Review*, 74(4), 1996, pp. 100 ff.

Y. Bakos, 'Information Links and Electronic Marketplaces: Implications of Interorganizational Information Systems in Vertical Markets', *Journal of Management Information Systems*, 8(2), 1991.

J. Black and O. Kharif, 'Second-Tier Portals: Going the Way of Go.com?', *Businessweek.com*, January 31 2001.

A.M. Brandenburger and B.J. Nalebuff, *Co-opetition*, Boston, MA., Harvard Business School Press, 1996.

A. Brown, 'Microsoft Corporation: Moving Beyond the Desktop', Deutsche Bank, November 3 2000.

CyberAtlas.com, 'National ISPs Still Kings of the ISP Hill', September 28 2000.

CyberAtlas.com, 'Wireless Aims for Widespread Appeal', February 13 2001.

M. Donegan, 'Contemplating Portal Strategies', *Telecommunications*, 34(2), 2000, pp. 48–52.

Economist, The, 'E-Commerce', February 26 2000, pp. 5–34.

D.C. Edelman, 'Patterns of Deconstruction: The Orchestrator', *Perspectives*, November 1998.

D. Einstein, 'Flat-Rate Net Access Finally Arrives on British Shores', *Forbes.com*, September 25 2000.

L. DiCarlo, 'IBM's Server Story', 11 May 2000.

J. Enrirquez, *As the Future Catches You: How Genomics and Other Forces are Changing Your Life, Work, Health and Wealth*, New York, Crown Business, 2001.

P.B. Evans, 'How Deconstruction Drives De-averaging', *Perspectives*, October 1998.

P. Evans and T. Wurster, *Blown to Bits. How the New Economics of Information Transform Strategy*, Boston, MSA., Harvard Business School Press, 1999.

J. Hagel III and M. Singer, 'Unbundling the Corporation', *Harvard Business Review*, March–April 1999, pp. 133–41.

A.C. Hax and D. Wilde II, *The Delta Project: Discovering New Sources of Profitability in a Networked Economy*, London, Palgrave 2001.

M.A. Hitt, J.E. Ricart and R.D. Nixon (eds), *Transactions Streams as Value Added: Sustainable Business Models on the Internet*, New York, John Wiley, 1998.

D. Lillington, 'Mobile but without Direction', *Wired.com*, September 21 2000.

T.W. Malone, J. Yates and R.I. Benjamin, 'Electronic Markets and Electronic Hierarchies', *Communications of the ACM*, June 1987.

Management Science, 'Reducing Buyer Search Costs: Implications for Electronic Marketplaces', 43(12), December 1997.

E.T. Penrose, *The Theory of the Growth of the Firms*, New York, John Wiley, 1958.

M. Porter, *Competitive Strategy: Techniques Analysing Industries and Competitors*, New York, Free Press, 1980.

M. Porter, *Competitive Advantage: Creating and Sustaining Superior Performance*, New York, Free Press, 1985.

J.E. Ricart, R.D. Nixon and M.A. Hitt, *Transaction Streams and Value Added: Sustainable Business Models on the Internet. New Managerial Mindsets: Organizational and Strategy Implementation*, New York, John Wiley, 1998.

J.E. Ricart, R. Subirana and J. Valor, *The Virtual Society*, IESE Publishing, 1997.

G. Saloner and A.M. Spence, *Creating and Capturing Value: Perspectives and Cases on Electronic Commerce*, John Wiley, 2002.

J. Schwartz, 'As Big PC Brother Watches, Users Encounter Frustration', *New York Time*, September 5 2001.

P. Seybold and R. Marshak, Customers.com: *How to Create a Profitable Business Strategy for the Internet and Beyond*, New York, Crown Business, 1998.

C. Shapiro and H.R. Varian, *Information Rules: A Strategic Guide to the Network Economy*, Boston, MA, Harvard Business School Press, 1999.

S. Sieber and J. Valor, 'Market Bundling Strategies in the Horizontal Portal Industry', *International Journal of Electronic Commerce*, Summer 2003.

C.W. Stern, 'The Deconstruction of Value Chains', *Perspectives*, September 1998.

B. Subirana, 'Zero Entry Barriers on a Computationally Complex World: Transaction Streams and the Complexity of the Digital Trade of Intangible Goods, *Journal of End-User Computing*, 1999.

B. Subirana, 'Readers Inn: The Future of Product and Services Distribution and the Transformation of the Publishing Industry', *7th European Conference on Information Systems*, 3, 23–25 June, Copenhagen 1999, pp. 964–83.

B. Subirana, 'The Transformation of the Publishing Industry and Sustainable Business Models: Readers Inn, Cases "A" and "B" ', *Journal of Information Technology Cases and Applications, Quarterly Journal*, 1(4), 1999, pp. 51–74.

B. Subirana and P. Carvajal, 'Transaction Streams: Theory and Examples Related to Confidence in Internet-Based Electronic Commerce', *Journal of Information Technology*, 15(1), March 2000, p. 3.

L. Thurow, 'Needed: A New System of Intellectual Property Rights', *Harvard Business Review*, 75(5), 1997, pp. 94–103.

B. Wallace, 'Web Hosting Heats Up', *Informationweek*, September 18 2000, pp. 22–5.